Biochemistry Research Trends

Life Sciences Research and Development

Biochemistry Research Trends

The Biochemical Guide to Toxins
David Aebisher, PhD, DSc (Editor)
Dorota Bartusik-Aebisher, PhD, DSc (Editor)
2023. ISBN: 979-8-89113-208-5 (Hardcover)
2023. ISBN: 979-8-89113-246-7 (eBook)

Airborne and Biological Monitoring to Assess Occupational Exposure to Isocyanates
Jimmy Hu, PhD (Editor)
2023. ISBN: 979-8-89113-010-4 (Softcover)
2023. ISBN: 979-8-89113-099-9 (eBook)

More information about this series can be found at
https://novapublishers.com/product-category/series/biochemistry-research-trends/

Life Sciences Research and Development

Woodlands: Ecology, Management and Threats
Ian D. Rotherham (Editor)
Salih Tunc Kaya, PhD (Editor)
2024. ISBN: 979-8-89113-248-1 (Hardcover)
2024. ISBN: 979-8-89113-307-5 (eBook)

J-Shaped Distributions and Their Applications
Mohammad Ahsanullah, PhD (Editor)
Mohammad Shakil (Editor)
2023. ISBN: 979-8-89113-013-5 (Softcover)
2023. ISBN: 979-8-89113-182-8 (eBook)

More information about this series can be found at
https://novapublishers.com/product-category/series/life-sciences-research-and-development/

Vasudeo Zambare
and Mohd. Fadhil Md. Din
Editors

Lipases and their Role in Health and Disease

Copyright © 2024 by Nova Science Publishers, Inc.

All rights reserved. No part of this book may be reproduced, stored in a retrieval system or transmitted in any form or by any means: electronic, electrostatic, magnetic, tape, mechanical photocopying, recording or otherwise without the written permission of the Publisher.

We have partnered with Copyright Clearance Center to make it easy for you to obtain permissions to reuse content from this publication. Please visit copyright.com and search by Title, ISBN, or ISSN.

For further questions about using the service on copyright.com, please contact:

Copyright Clearance Center
Phone: +1-(978) 750-8400   Fax: +1-(978) 750-4470   E-mail: info@copyright.com

NOTICE TO THE READER

The Publisher has taken reasonable care in the preparation of this book but makes no expressed or implied warranty of any kind and assumes no responsibility for any errors or omissions. No liability is assumed for incidental or consequential damages in connection with or arising out of information contained in this book. The Publisher shall not be liable for any special, consequential, or exemplary damages resulting, in whole or in part, from the readers' use of, or reliance upon, this material. Any parts of this book based on government reports are so indicated and copyright is claimed for those parts to the extent applicable to compilations of such works.

Independent verification should be sought for any data, advice or recommendations contained in this book. In addition, no responsibility is assumed by the Publisher for any injury and/or damage to persons or property arising from any methods, products, instructions, ideas or otherwise contained in this publication.

This publication is designed to provide accurate and authoritative information with regards to the subject matter covered herein. It is sold with the clear understanding that the Publisher is not engaged in rendering legal or any other professional services. If legal or any other expert assistance is required, the services of a competent person should be sought. FROM A DECLARATION OF PARTICIPANTS JOINTLY ADOPTED BY A COMMITTEE OF THE AMERICAN BAR ASSOCIATION AND A COMMITTEE OF PUBLISHERS.

Library of Congress Cataloging-in-Publication Data

ISBN: 979-8-89113-628-1 (softcover)
ISBN: 979-8-89113-674-8 (e-book)

Published by Nova Science Publishers, Inc. † New York

In loving memory of

**Mrs. Janabai Pandharinath Zambare
(07 Jul 1950 - 29 Jun 2023)**

This book, "Lipases and their Role in Health and Disease," is dedicated to the cherished memory of my beloved mother, Mrs. Janabai Pandharinath Zambare. Her journey through life was marked by resilience, warmth, and an unwavering spirit, making an indelible mark on those fortunate enough to know her.

Her struggle with high body fat was a challenge she faced with grace and tenacity, and it is in her honor that this work explores the intricate world of lipases. As we unravel the profound roles these enzymes play in health and disease, it is a testament to her enduring spirit and the hope that future innovations in lipase research may contribute to alleviating obesity-related challenges for generations to come. Her enduring encouragement and belief in the pursuit of knowledge have been a source of inspiration. This book is a testament to the values she instilled — a commitment to excellence, a passion for learning, and the courage to explore new frontiers.

Though she is no longer with us, her legacy lives on through the ideas, principles, and the enduring love she shared. As we delve into the intricate world of lipases, it is a reflection of the curiosity and intellectual curiosity she cultivated in me.

May this work stand as a humble homage to a remarkable woman, a loving mother, and a beacon of strength. Her memory will forever be etched in our hearts.

With love.
Dr. Vasudeo Zambare

Contents

Preface		ix
Acknowledgments		xi
Abbreviations		xiii
Chapter 1	**Overview on Lipases and Their Role in Health and Disease** 1	
	Vasudeo Zambare and Mohd. Fadhil Md. Din	
Chapter 2	**Lipase: Beyond Digestion - Exploring Health Benefits, Metabolic Influence, and Therapeutic Potential** 23	
	Sandhya Mulchandani and Pratik Kale	
Chapter 3	**Probiotic Lipases: Versatile Catalysts for Sustainable Industries and Therapeutic Innovations** 65	
	Saraswathy Nagendran and Neha Mudaliar	
Chapter 4	**Bio-Surfactants: Green Synthesis Using Lipases and Role in Human Health** 87	
	Benu Arora	
Chapter 5	**Lipases: Guardians of Energy Metabolism and Beyond - Unveiling Therapeutic, Physiological, and Functional Frontiers** 121	
	Gagandeep Kaur and Vikas Sharma	
Chapter 6	**Lipases in Diseased Conditions – Impact on Humans, Animals and Birds** 141	
	Sourabh Patil and Tushar Borse	

About the Editors 163
List of Contributors 167
Index 169

Preface

In the vast tapestry of human health and disease, lipases emerge as silent orchestrators, weaving together the intricate biochemistry that underlies our physiological processes. As we present this groundbreaking work, "Lipases and their Role in Health and Disease," we invite readers into a realm where enzymes transcend their conventional roles, revealing a profound impact on the delicate balance of our well-being.

The journey unfolds with Chapter 1, "Overview on Lipases and Their Role in Health and Disease," authored by Vasudeo Zambare and Mohd. Fadhil Md. Din. This chapter serves as the cornerstone, embarking on a meticulous exploration of lipase biochemistry. From unravelling molecular structures to scrutinizing the distinctive roles of lipases in lipid metabolism, this study unveils the pivotal connection between lipases and digestion. As we delve deeper, the chapter sheds light on the multifaceted picture of lipases in health, their contribution to nutrient absorption, cellular function, and immune system modulation. However, the narrative doesn't shy away from the challenges, extending to lipase-related disorders and the promising therapeutic applications that lie ahead.

In Chapter 2, "Lipase: Beyond Digestion - Exploring Health Benefits, Metabolic Influence, and Therapeutic Potential," Sandhya Mulchandani and Pratik Kale take readers on a journey beyond the digestive realm of lipases. From their crucial role in fat breakdown during rest and activity to applications in diverse industries, this chapter unravels the therapeutic, physiological, and functional frontiers of lipases. The versatility of these enzymes, from lipid substrate digestion to membrane stability and lipid rafts' creation, is vividly portrayed.

Chapter 3, "Probiotic Lipases: Versatile Catalysts for Sustainable Industries and Therapeutic Innovations," penned by Saraswathy Nagendran and Neha Mudaliar, addresses the global issue of obesity. Lipases, heralded as essential enzymes, prove to be not only important for future aspects but also offer sustainability in various industries. The inhibition of pancreatic lipases using probiotic food products is explored, highlighting their role in enhancing anti-obesity properties and provoking weight loss through microbial fermentation.

As we transition to Chapter 4, "Bio-surfactants: Green Synthesis using Lipases and Role in Human Health," Benu Arora takes canter stage. This chapter introduces the environmentally friendly realm of bio-surfactants, emphasizing glycolipids' eco-friendly synthesis. The latter part of the chapter delves into biomedical applications, underscoring the antimicrobial and antibiofilm activities of glycolipids as pivotal elements in preventing microbial infections and serving as drug delivery and anti-cancer agents.

Chapter 5, "Lipases: Guardians of Energy Metabolism and Beyond- Unveiling Therapeutic, Physiological and Functional Frontiers," authored by Gagandeep Kaur and Vikas Sharma, delves into the multifaceted roles of lipases. From their involvement in fat breakdown during rest and activity to their applications in various industries, readers discover the guardianship of energy metabolism by these versatile enzymes. This chapter underscores the therapeutic, physiological, and functional frontiers of lipases, emphasizing their pivotal role in lipid substrate digestion, membrane stability, and the creation of lipid rafts.

Chapter 6, "Lipases in Diseased Conditions – Impact on Humans, Animals and Birds," authored by Sourabh Patil and Tushar Borse, marks a significant pivot. This chapter navigates through the ubiquity of cellular enzyme lipases, exploring their hydrolytic roles in converting triglycerides into fatty acids and glycerol. The varied substrates, tissue-specific expressions, and pivotal roles in maintaining health are vividly illuminated, alongside their implications in diverse diseases.

As we present this collaborative endeavour, we hope to unravel the captivating saga of lipases in health and disease. From molecular intricacies to clinical applications, each chapter contributes a unique thread to the rich tapestry of enzymatic research. Our journey through these pages promises not only a deeper understanding of lipase biology but also beckons towards future avenues of precision therapies and innovative treatments. Join us on this enlightening odyssey into the world of lipases, where science meets the intricate dance of life.

Dr. Vasudeo Zambare
Professor Dr. Mahd. Fadhil Md. Din

Acknowledgments

We extend our heartfelt gratitude to the dedicated contributors and collaborators who made "Lipases and their Role in Health and Disease" a reality. This monumental work would not have been possible without the unwavering commitment and expertise of our esteemed authors.

Special thanks to our authors Sandhya Mulchandani, Pratik Kale, Saraswathy Nagendran, Neha Mudaliar, Benu Arora, Gagandeep Kaur, Vikas Sharma, Sourabh Patil, and Tushar Borse for their scholarly insights and comprehensive chapters. Your contributions have added depth and diversity to our exploration of lipases, expanding our understanding of their roles in health and disease.

Mr. Abhay Kainya, Director of Balaji Enzyme and Chemical Pvt Ltd, India, deserves special acknowledgment for his invaluable contributions. His visionary leadership and commitment to advancing enzymatic research have been a guiding force throughout the development of this book.

We express our sincere and deepest appreciation to Professor Dr. Zainura Zainon Noor, Director of the Center for Environmental Sustainability and Water Security (IPASA) and Dr. Suhaimi bin Abu Bakar, Dean of the Department of Water and Environmental Engineering, School of Civil Engineering at Universiti Teknologi Malaysia, Bahru, Malaysia. We are truly grateful for their dedication, expertise, and encouragement.

A sincere thank you to the publishers and Mr. Thomas Mazzaferro for their vision and commitment to advancing scientific knowledge. Your support has provided a platform for sharing groundbreaking research and fostering dialogue within the scientific community.

We express our gratitude to the editorial management team for their meticulous efforts in coordinating the diverse contributions and ensuring the cohesiveness of this book. Their dedication to maintaining the highest standards of academic rigor has been instrumental in shaping this pioneering work.

This collaborative effort embodies the spirit of intellectual curiosity and the pursuit of knowledge. We appreciate everyone who played a role, no matter how small, in bringing "Lipases and their Role in Health and Disease" to fruition.

Thank you for being part of this transformative journey.

Sincerely,

Dr. Vasudeo Zambare
Professor Dr. Mahd. Fadhil Md. Din

Abbreviations

2-AG	2-Arachidonoylglycerol
2M2B	2-Methyl-2-Butanol
AME	Apparent Metabolizable Energy
ATGL	Adipose Triglyceride Lipase
ATP	Adenosine Ttriphosphate
BMI	Body Mass Index
CAC	Cancer-Associated Cachexia
CAGR	Compound Annual Growth Rate
CALB	Candida antarctica Lipase B
CBR	Cannabinoid Receptor
CCK	Cholecystokinin
CE	Cholesteryl Ester
CESD	Cholesteryl Ester Storage Disease
CF	Cystic Ffibrosis
CFTR	Cystic Fibrosis Transmembrane Regulator
CHD	Coronary Heart Disease
CIA	Collagen-Induced Arthritis
CLEA	Cross-linked Enzyme Aggregates
CLPCMC	Cross-linked Protein Coated Micro-Crystals
CMC	Critical Micelle Concentration (CMC)
CNA	Coefficient of Nitrogen Absorption
CP	Chronic Pancreatitis
CRP	C-reactive Protein
CT	Computed Tomography
CVD	Cardiovascular Disease
DAG	Diacylglycerols
DAGL	Diacylglycerol Lipase
DES	Deep Eutectic Solvents
DMSO	Dimethyl Sulphoxide
EC	Endothelial Cell
EL	Endothelial Lipase

EPI	Exocrine Pancreatic Insufficiency
EPROS	Enzyme Precipitated and Rinsed with Organic Solvent
ERCP	Endoscopic Retrograde Cholangiopancreatography
ERT	Enzyme Replacement Therapy
FA	Fatty Acids
FAC	Fat Absorption Coefficient
FAVE	Fatty Acid Vinyl Ester
FC	Free Cholesterol
FDA	Food and Drug Administration
FE-1	Faecal Elastase-1
FFA	Free Fatty Acids
GDDR	1,2-o-dilauryl-rac-glycero-3-glutaric acid-(6'-methylresorufin) ester
GI	Gastrointestinal
GPIHBP1	Glycosyl Phosphatidyl Inositol-Anchored High-Density Lipoprotein-Binding Protein 1
GRAS	Generally Regarded As Safe
HABP	High Activity Biocatalyst Preparation
HDL	High-Density lipoproteins
HL	Hepatic Lipase
HLB	Hydrophilic-Lipophilic Balance
HPL	Human Pancreatic Lipase
HSL	Hormone-Sensitive Lipase
HSPG	Heparan Sulfate Proteoglycan
IBD	Inflammatory Bowel Disease
IDL	Intermediate-Density Lipoprotein
IL	Ionic Liquid
IPASA	Center for Environmental Sustainability and Water Security
IU	international units
LAB	Lactic Acid Bacteria
LAL	Lysosomal Acid Lipase
LD	Lipid Droplets
LDL	Low-Density Lipoprotein
LIPA	Lysosomal Acid Lipase
Lp(a)	Lipoprotein A
LPL	Lipoprotein Lipase
LST	Lipase Supplementation Therapy
MAG	Monoacylglycerols
MAGL	Monoacylglycerol Lipase

MEL	Mannosylerythritol Lipid
MG	Monoacylglycerol
MGGM	(±) methyl trans-3(4-methoxyphenyl) Glycidate
MGL	Monoglyceride Lipase
MRI	Resonance Imaging
NLSD-M	Neutral Lipid Storage Disease-Associated Myopathy
NSAID	Non-steroidal Anti-inflammatory Drugs
PAMP	Pathogen-Associated Molecular Patterns
PCMC	Protein Coated Micro-Crystals
PERT	Pancreatic Enzyme Replacement Therapy
PI	Pancreatic Insufficiency
PL	Pancreatic Lipase
PLA2	Phospholipase A2
PLI	Pancreatic Lipase Immunoreactivity
RCT	Reverse Cholesterol Transport
SCFA	Short-Chain Fatty Acids
sdLDL	Small, Dense LDL
SFAE	Sugar Fatty Acid Esters
SNP	Single Nucleotide Polymorphism
TG	Triglycerides
TGF	Transforming Growth Factor
THL	Tetrahydrolipstatin
TNF	Tumour Necrosis Factor
TTS	Transdermal Therapeutic Systems
VLDL	Very Low-Density Lipoproteins
WHO	World Health Organization

Chapter 1

Overview on Lipases and Their Role in Health and Disease

Vasudeo Zambare[1,2,*] and Mohd. Fadhil Md. Din[2]

[1]R&D Department, Balaji Enzyme and Chemical Pvt. Ltd., Andheri (E), Mumbai, India
[2]Center for Environmental Sustainability and Water Security (IPASA) and Department of Water and Environmental Engineering, School of Civil Engineering, Universiti Teknologi Malaysia, Bahru, Malaysia

Abstract

This comprehensive exploration delves into the intricate world of lipases and their profound influence on human health and disease. The journey begins with a meticulous investigation of the biochemistry of lipases, unraveling their molecular structure, classification, and enzymatic mechanisms. Types of lipases, including pancreatic lipase, gastric lipase, lipoprotein lipase, and adipose tissue lipase, are scrutinized for their distinctive roles in lipid metabolism. The pivotal role of lipases in digestion emerges as a focal point, emphasizing lipid hydrolysis, absorption of fatty acids, and the formation of micelles as critical processes in nutrient assimilation. The connection between lipases and metabolism is explored, elucidating their regulation, impact on energy homeostasis, and influence on insulin sensitivity. Lipases in health unveil a multifaceted picture, contributing to nutrient absorption, cellular function, and immune system modulation. However, lipases are not immune to disease, and the exploration extends to lipase deficiency, pancreatitis, and metabolic disorders. Diagnostic applications, encompassing lipase level measurement, imaging techniques, and

---

[*] Corresponding Author's Email: vasuzambare@gmail.com.

In: Lipases and their Role in Health and Disease
Editors: Vasudeo Zambare and Mohd. Fadhil Md. Din
ISBN: 979-8-89113-628-1
© 2024 Nova Science Publishers, Inc.

biomarkers, provide crucial tools for identifying and understanding lipase-related disorders. Therapeutic implications, including enzyme replacement therapy, lipase inhibitors, and potential future developments, showcase practical applications in managing lipase-associated conditions. In conclusion, this study encapsulates the richness of lipase research, from its molecular foundations to its clinical applications. The abstract highlights the interconnectedness of lipases with diverse physiological processes, emphasizing their significance in health and the challenges they pose in disease. As we delve into future research, the prospects for precision therapies and innovative treatments beckon, promising a deeper understanding of lipase biology and novel strategies for personalized interventions.

Keywords: lipase, digestion, health, disease, metabolc disorder, diagnostic

Introduction

Lipases, a crucial group of enzymes, play a fundamental role in the intricate processes of lipid metabolism within the human body (Sahu and Birner-Gruenberger, 2013). These enzymes are pivotal in the hydrolysis of triglycerides, breaking them down into fatty acids and glycerol. The significance of lipases extends beyond their involvement in digestion; they contribute substantially to cellular functions, energy homeostasis, and overall health maintenance (Chandra et al., 2020; Grabner et al., 2021). Understanding the biochemistry of lipases is essential for unraveling their diverse roles. Lipases exhibit a complex molecular structure, and their classification is based on their specific functions within the body. From pancreatic lipases responsible for digestive processes to lipoprotein lipases involved in lipid transport, these enzymes operate in a finely tuned enzymatic mechanism (Zhu et al., 2021). This exploration delves into the various types of lipases, emphasizing their distinct functions in different physiological contexts. Pancreatic lipases, gastric lipases, lipoprotein lipases, and adipose tissue lipases each contribute uniquely to lipid metabolism and overall health (Zechner et al., 2012). Beyond their roles in digestion, lipases significantly impact metabolic processes. The regulation of lipases is intricately linked to energy homeostasis and insulin sensitivity, making them key players in metabolic health (Holm et al., 2000). This section will elucidate the intricate connections between lipases and the broader metabolic framework.

As we embark on this exploration of lipases, it becomes evident that these enzymes are not only indispensable for nutrient absorption but also crucial for maintaining cellular function and modulating the immune system. The delicate balance of lipase activity contributes to the overall well-being of an individual. However, disruptions in lipase function can lead to various health challenges. Lipase deficiency, pancreatitis, and metabolic disorders underscore the importance of these enzymes in preventing disease states (Balasubramanian et al., 2023). Unraveling the mechanisms underlying these conditions is vital for developing diagnostic tools and therapeutic interventions. This chapter exploring the diagnostic applications of lipase research, shedding light on how measurements of lipase levels and advanced imaging techniques contribute to our understanding of health and disease.

Furthermore, this chapter delve into the potential therapeutic implications, including enzyme replacement therapy, lipase inhibitors, and emerging possibilities on the horizon. The navigation of multifaceted landscape of lipases and their roles in health and disease, the insights gained have far-reaching implications for both current medical practices and the future of healthcare.

Biochemistry of Lipases

Lipases, critical components in lipid metabolism, exhibit a sophisticated biochemistry that governs their functions within the human body. This section will elucidate the intricate details of lipase biochemistry, focusing on their molecular structure, classification, and the enzymatic mechanisms that underscore their catalytic activity (Pirahanchi et al., 2023).

Molecular Structure

The molecular structure of lipases is a testament to their versatility and adaptability. Lipases are generally characterized by a hydrophobic core surrounded by hydrophilic regions, facilitating their interaction with both water and lipid substrates. This amphipathic nature enables lipases to navigate the aqueous environment of biological systems while catalysing reactions at the lipid-water interface. The active site of a lipase, where catalysis occurs, is composed of amino acid residues that create a specific three-dimensional pocket. This pocket accommodates the hydrophobic tail of lipid substrates, allowing for precise and selective enzymatic activity. The molecular architecture of lipases thus embodies a finely tuned structure-function relationship that is central to their biological roles (Lotti and Alberghina, 2007).

Classification of Lipases

Lipases are a diverse group of enzymes, classified based on their source of origin and specific functions. One common classification distinguishes between pancreatic lipases, gastric lipases, lipoprotein lipases, and adipose tissue lipases (Zechner et al., 2012; Sahu and Birner-Gruenberger, 2013).

- *Pancreatic Lipases:* Primarily involved in the digestion of dietary fats, pancreatic lipases are secreted by the pancreas into the small intestine. Their role in breaking down triglycerides into absorbable fatty acids and glycerol is essential for nutrient absorption.
- *Gastric Lipases:* Secreted in the stomach, gastric lipases initiate lipid digestion in the acidic environment of the gastric lumen. While their contribution is minor compared to pancreatic lipases, they play a crucial role in the initial stages of lipid breakdown.
- *Lipoprotein Lipases:* Found on the surface of cells, lipoprotein lipases catalyze the hydrolysis of triglycerides within circulating lipoproteins. This process releases fatty acids for cellular uptake and utilization.
- *Adipose Tissue Lipases:* Enzymes in adipose tissue, such as hormone-sensitive lipase, contribute to the mobilization of stored fats during times of energy demand. Their role in lipolysis influences overall energy homeostasis.

Enzymatic Mechanism

The enzymatic mechanism of lipases involves a series of precisely orchestrated steps. Lipases are characterized by their ability to hydrolyze ester bonds, cleaving the bonds between glycerol and fatty acids in triglycerides (Pirahanchi et al., 2023). This hydrolysis occurs at the lipid-water interface, facilitated by the amphipathic nature of lipase molecules. The catalytic triad, consisting of serine, histidine, and aspartate or glutamate residues, is a hallmark of lipase enzymatic activity. Serine acts as the nucleophile, initiating the attack on the ester bond in the lipid substrate. The subsequent steps involve acylation and deacylation, ultimately resulting in the release of fatty acids and glycerol. The specificity of lipases for particular substrates, combined with their ability to function in diverse biological environments, underscores the elegance of their enzymatic mechanisms. Understanding these intricate processes is crucial for unravelling the broader roles of lipases in health and disease.

Thus, the biochemistry of lipases, encompassing their molecular structure, classification, and enzymatic mechanisms, provides a foundational understanding essential for exploring their multifaceted contributions to physiological processes.

Types of Lipases

Lipases, a diverse group of enzymes, exhibit specialized functions depending on their origin and localization within the body. This section will delve into the distinct types of lipases, namely Pancreatic Lipase, Gastric Lipase, Lipoprotein Lipase, and Adipose Tissue Lipase, elucidating their unique roles in lipid metabolism.

Pancreatic Lipase

- *Origin and Secretion:* Pancreatic lipase, a key player in lipid digestion, is primarily synthesized and secreted by the pancreas. It is released into the small intestine in response to the presence of dietary fats (Carrière et al., 1994).
- *Function:* Pancreatic lipase plays a central role in the hydrolysis of triglycerides present in ingested food. It cleaves the ester bonds between fatty acids and glycerol, producing free fatty acids and monoglycerides. This process is essential for the absorption of dietary lipids in the small intestine (Carrière et al., 1994).
- *Regulation:* The secretion and activity of pancreatic lipase are tightly regulated to match dietary lipid intake. Hormonal signals, such as cholecystokinin (CCK) released in response to the presence of fats, stimulate the release of pancreatic lipase (Brownlee et al., 2010).

Gastric Lipase

- *Origin and Secretion:* Gastric lipase is primarily secreted by the gastric chief cells in the stomach lining. While its contribution to overall lipid digestion is relatively minor compared to pancreatic lipase, it initiates the breakdown of dietary fats in the stomach (Tomasik et al., 2013).
- *Function:* Gastric lipase is active in the acidic environment of the stomach. It catalyses the hydrolysis of triglycerides into diglycerides and free fatty acids. This initial step in lipid digestion prepares the substrate for further processing in the small intestine (Miled et al., 2000).
- *Role in Early Digestion:* Gastric lipase acts as an essential component in the early stages of lipid digestion, complementing the subsequent actions of pancreatic lipase in the small intestine (Sassene et al., 2016).

Lipoprotein Lipase

- *Localization:* Lipoprotein lipase is found on the surface of cells, particularly in tissues that require fatty acids for energy or storage, such as adipose tissue and muscle (Wang and Eckel, 2009).
- *Function:* Lipoprotein lipase plays a crucial role in the hydrolysis of triglycerides present in circulating lipoproteins. It releases fatty acids, which can then be taken up by cells for energy production or storage. This enzyme is instrumental in regulating lipid distribution and utilization throughout the body (Pirahanchi et al., 2023).
- *Regulation:* The activity of lipoprotein lipase is regulated by various factors, including insulin, which enhances its function. This enzyme contributes to the balance of lipid metabolism and energy homeostasis (Kersten, 2014).

Adipose Tissue Lipase

- *Variants:* Adipose tissue lipases encompass various enzymes, with adipose triglyceride lipase (ATGL), hormone-sensitive lipase (HSL), and monoglyceride lipase (MGL) (Zechner et al., 2009).
- *Function:* Adipose tissue lipases, particularly hormone-sensitive lipase, are involved in the hydrolysis of triglycerides stored in adipocytes. This process releases fatty acids and glycerol during periods of increased energy demand, such as fasting or exercise (Reynisdottir et al., 1997).
- *Regulation:* Hormonal signals, including catecholamines and insulin, regulate the activity of adipose tissue lipases. This regulation ensures a dynamic response to the body's energy requirements (Li and Sun, 2018; De Koster et al., 2018).

Understanding the roles of these diverse lipases—pancreatic lipase, gastric lipase, lipoprotein lipase, and adipose tissue lipase—provides a comprehensive view of their contributions to lipid metabolism, nutrient absorption, and overall physiological homeostasis.

Role of Lipases in Digestion

Lipases play a pivotal role in the digestive process, orchestrating the breakdown of dietary fats and facilitating their absorption in the gastrointestinal tract. This section explores the specific contributions of lipases to lipid hydrolysis, absorption of fatty acids, and the formation of micelles, highlighting their significance in nutrient assimilation.

Lipid Hydrolysis

The primary role of lipases in digestion lies in catalyzing the hydrolysis of triglycerides, the predominant form of dietary fats. Triglycerides consist of three fatty acid chains esterified to a glycerol backbone. Lipases, particularly pancreatic lipase and gastric lipase, act on these ester bonds, cleaving them to release free fatty acids and glycerol (Cerk et al., 2018). Pancreatic lipase, secreted by the pancreas into the small intestine, continues the process of lipid hydrolysis initiated by gastric lipase in the stomach. These enzymatic activities collectively transform complex triglycerides into simpler components, facilitating their subsequent absorption (Zhu et al., 2021).

Absorption of Fatty Acids

The products of lipase-mediated triglyceride hydrolysis, namely free fatty acids and monoglycerides, face a challenge in traversing the watery environment of the small intestine. To overcome this hurdle, bile salts released by the gallbladder emulsify the lipid breakdown products, creating structures known as micelles. Micelles serve as carriers for the solubilized lipids, ensuring their efficient transport to the surface of enterocytes lining the small intestine. This enables the absorption of free fatty acids and monoglycerides into the enterocytes for further processing (Garrett and Young, 1975). Once inside the enterocytes, the absorbed lipids are reassembled into triglycerides. These newly formed triglycerides are then incorporated into chylomicrons, which are large lipoprotein particles that transport lipids through the lymphatic system and eventually into the bloodstream.

Micelle Formation

Bile salts, critical components of bile released by the liver and stored in the gallbladder, play a key role in the formation of micelles (Cheng et al., 2020). These amphipathic molecules surround and solubilize the hydrophobic products of lipid digestion, creating micellar structures that are more amenable to transport in the aqueous environment of the small intestine. Micelles present a significantly increased surface area for lipase activity, promoting the efficient hydrolysis of lipids into absorbable components. This process ensures that lipids, which are inherently hydrophobic, can be effectively processed and absorbed in the water-rich environment of the digestive tract. Understanding the intricate role of lipases in lipid hydrolysis, absorption of fatty acids, and micelle formation provides crucial insights into the mechanisms governing the digestion and assimilation of dietary fats. This orchestrated interplay of enzymes and molecular structures is fundamental to the efficient utilization of lipids for energy and essential cellular functions.

Lipases and Metabolism

Lipases are integral players in the complex network of metabolic processes, influencing energy balance and cellular homeostasis. This section delves into the regulation of lipases, their role in maintaining energy homeostasis, and their intricate connection with insulin sensitivity.

Lipase Regulation

- *Hormonal Control:* The activity of lipases is tightly regulated by hormonal signals, ensuring a dynamic response to the body's metabolic needs. Hormones such as glucagon and adrenaline stimulate lipase activity, promoting the breakdown of stored triglycerides into fatty acids and glycerol (Kraemer and Shen, 2002).
- *Insulin Modulation:* Conversely, insulin, released by the pancreas in response to elevated blood glucose levels, exerts an inhibitory effect on lipase activity. This insulin-mediated suppression serves to conserve energy during times of nutrient abundance, directing cells to prioritize glucose utilization over lipid breakdown (Monika et al., 2012).
- *Substrate Availability:* Lipase regulation is also influenced by substrate availability. Elevated levels of circulating triglycerides, as seen in the postprandial state, stimulate lipase activity to facilitate the absorption and utilization of dietary lipids (Liu et al., 2020).

Energy Homeostasis

Lipases play a crucial role in maintaining energy homeostasis by regulating the availability of fatty acids as a fuel source. During periods of energy demand, lipases mobilize stored triglycerides in adipose tissue, releasing fatty acids that can be utilized by tissues such as muscle for energy production (Althaher, 2022). Adipose tissue lipases, including hormone-sensitive lipase, contribute significantly to the dynamic storage and release of energy. The controlled release of fatty acids from adipose tissue provides a sustained energy source during fasting or physical activity (Kulminskaya and Oberer, 2020).

Lipase and Insulin Sensitivity

Insulin, a key regulator of glucose metabolism, also exerts profound effects on lipid metabolism. In addition to its inhibitory action on lipase activity, insulin enhances the uptake of circulating fatty acids by cells, promoting their storage as triglycerides (Vargas and Joy, 2023). Insulin sensitivity, the responsiveness of cells to insulin, plays a crucial role in lipid utilization. Individuals with enhanced insulin sensitivity exhibit efficient glucose uptake and reduced reliance on lipolysis for energy, contributing to overall metabolic health (Zechner et al., 2012). Dysregulation of lipase activity and impaired insulin sensitivity are associated with metabolic disorders such as obesity and type 2 diabetes. Understanding the intricate interplay between lipases and insulin sensitivity is essential for unraveling the mechanisms underlying these conditions and developing targeted therapeutic interventions (Blaak, 2003).

Overall, the regulation of lipases, their contribution to energy homeostasis, and their connection with insulin sensitivity form an intricate nexus within the broader landscape of metabolism. Unraveling the nuances of these interactions provides valuable insights into the physiological mechanisms governing metabolic health and the potential avenues for therapeutic interventions in metabolic disorders.

Lipases in Health

Lipases, key enzymes in lipid metabolism, play a multifaceted role in maintaining health by contributing to nutrient absorption, cellular function, and immune system modulation. This section explores the diverse ways in which lipases contribute to overall well-being (Loli et al., 2015).

Nutrient Absorption

Lipases, particularly pancreatic and gastric lipases, are instrumental in breaking down dietary fats into absorbable components. Through the hydrolysis of triglycerides, lipases generate free fatty acids and monoglycerides. This process is essential for the efficient absorption of lipids in the small intestine. Additionally, lipases contribute to the formation of micelles, enhancing the solubility of lipids and facilitating their transport across the enterocyte membrane (Baurer et al., 2005). Lipases also play a crucial role in the absorption of fat-soluble vitamins, such as A, D, E, and K (Loli et al., 2015). These vitamins are co-absorbed with dietary fats, and the activity of lipases ensures their incorporation into chylomicrons for systemic distribution (Reddy and Jailal, 2023).

Cellular Function

The liberated fatty acids from lipase-mediated hydrolysis serve as a vital energy source for various cell types. Cellular uptake of these fatty acids supports energy production through processes such as beta-oxidation in mitochondria (Li and Zhang, 2019). This energy contribution is particularly significant in tissues with high metabolic demands, such as muscle cells. Lipids derived from dietary fats, processed with the help of lipases, contribute to the structural integrity of cell membranes. Phospholipids and cholesterol, products of lipid metabolism, play essential roles in maintaining cell membrane fluidity and function.

Immune System Modulation

Lipids generated by lipase activity also serve as precursors for bioactive molecules with immunomodulatory properties (Flaherty et al., 2019). Certain fatty acids, such as omega-3 polyunsaturated fatty acids, have anti-inflammatory effects and play a role in regulating immune responses. Lipases contribute to the release of these bioactive lipids, influencing immune system function (Habib et al., 2019). Lipids are involved in signaling pathways that modulate immune cell functions. Lipase-generated lipid mediators can influence the activities of immune cells, including macrophages and T cells, in response to infections and inflammatory challenges (Hubler and Kennedy, 2016). Lipids are essential for the development and function of immune cells. Lipase-mediated lipid metabolism contributes to the synthesis of lipids necessary for immune cell maturation and activity (Howie et al., 2018).

Based on the multifaceted contributions of lipases in nutrient absorption, cellular function, and immune system modulation provides insights into their essential roles in maintaining health. The delicate balance maintained by lipases in these processes underscores their significance in supporting the overall well-being of the organism.

Lipases in Disease

While lipases are crucial for maintaining physiological processes, their dysregulation or deficiency can contribute to various diseases. This section explores the impact of lipases in disease, specifically focusing on lipase deficiency, pancreatitis, and metabolic disorders (Subramoni et al., 2010).

Lipase Deficiency

Lipase deficiency can result from both genetic factors and acquired conditions. Genetic disorders, such as congenital lipase deficiency, lead to a lack or malfunction of lipase enzymes (Ghishan et al., 1984). Acquired causes may include conditions that impair the production or activity of lipases, such as certain pancreatic disorders or diseases affecting the small intestine. Lipase deficiency severely impairs the hydrolysis of dietary fats, hindering the absorption of essential fatty acids and fat-soluble vitamins. This can lead to malnutrition, deficiencies in critical nutrients, and subsequent health complications. Lipase deficiency is often associated with symptoms such as steatorrhea (excessive fat in the feces), abdominal discomfort, and failure to thrive, particularly in infants and young children (Balasubramanian et al., 2023).

Pancreatitis

Pancreatitis, inflammation of the pancreas, is a condition where lipases play a central role (Ater and Koch, 2021). The pancreas, home to pancreatic lipases, is susceptible to self-digestion during inflammatory processes. In pancreatitis, lipases are inappropriately activated within the pancreas, leading to the digestion of pancreatic tissue (De Oliveira et al., 2020). The release of lipases into the pancreatic tissue causes damage and inflammation. This can result in severe abdominal pain, nausea, vomiting, and, in severe cases, systemic complications (Mohy-ud-din and Morrissey, 2023). Elevated levels of circulating lipases are often observed in blood tests, aiding in the diagnosis of pancreatitis. Chronic pancreatitis, characterized by persistent inflammation, can lead to long-term damage to the pancreas. This can result in the loss of pancreatic function, including lipase secretion, contributing to malabsorption of nutrients and the development of nutritional deficiencies.

Metabolic Disorders

- *Lipase Dysregulation in Metabolic Syndrome:* Lipase dysregulation is implicated in metabolic disorders such as obesity and metabolic syndrome. In conditions of excess adiposity, lipase activity may be altered, leading to an imbalance in lipid metabolism (Nagarajan et al., 2022).
- *Impact on Insulin Sensitivity:* Altered lipase activity can contribute to insulin resistance, a key feature of metabolic syndrome and type 2 diabetes. Dysregulated lipase activity may lead to an imbalance in the release and storage of fatty acids, influencing insulin sensitivity in target tissues.
- *Associations with Cardiovascular Risk:* Imbalances in lipase activity and lipid metabolism in metabolic disorders are associated with an increased risk of cardiovascular diseases. Dyslipidemia, characterized by abnormal lipid profiles, is a common feature, highlighting the intricate interplay between lipases and metabolic health (Zambon et al., 2003).

Understanding the role of lipases in diseases such as lipase deficiency, pancreatitis, and metabolic disorders provides crucial insights into potential therapeutic targets and interventions aimed at restoring lipid homeostasis and mitigating associated health risks.

Diagnostic Applications

Diagnostic applications related to lipases are crucial for assessing various health conditions, ranging from pancreatic disorders to metabolic diseases. This section explores the diverse approaches, including lipase level measurement, imaging techniques, and biomarkers, utilized in diagnostics (Qiao et al., 2020).

Lipase Level Measurement

- *Laboratory Assessments:* Lipase level measurement in blood is a fundamental diagnostic tool, particularly in assessing pancreatic

health. Elevated levels of circulating lipases, such as pancreatic lipase, can indicate pancreatic injury or inflammation, as observed in conditions like pancreatitis.
- *Diagnostic Significance:* Lipase level measurement is particularly valuable in distinguishing pancreatitis from other causes of abdominal pain and discomfort. Persistently elevated lipase levels help confirm the diagnosis and monitor the progression of the condition.
- *Clinical Evaluation:* Lipase levels are often evaluated alongside amylase levels, another pancreatic enzyme. The ratio of lipase to amylase can provide additional diagnostic information and aid in differentiating between acute pancreatitis and other conditions.

Imaging Techniques (Lippi et al., 2012)

- *Computed Tomography (CT) Scans:* Imaging techniques, such as CT scans, are instrumental in visualizing the pancreas and surrounding structures. In cases of pancreatitis or pancreatic tumours, CT scans can reveal structural abnormalities, inflammation, or the presence of cysts.
- *Magnetic Resonance Imaging (MRI):* MRI is another non-invasive imaging modality used to assess pancreatic health. It provides detailed images of the pancreas, aiding in the diagnosis and characterization of various pancreatic disorders.
- *Endoscopic Retrograde Cholangiopancreatography (ERCP):* ERCP combines endoscopy and fluoroscopy to visualize the pancreatic and bile ducts. It is particularly useful for diagnosing conditions such as pancreatitis, pancreatic tumours, or obstructions in the ducts.

Biomarkers for Disease (Meher et al., 2015; Jawed et al., 2019)

- *Circulating Lipid Profiles:* Lipids, products of lipase activity, can serve as biomarkers for assessing metabolic health. Abnormal lipid profiles, including elevated triglycerides and cholesterol levels, may indicate dyslipidemia and contribute to the diagnosis of metabolic disorders.

- *Inflammatory Markers:* Inflammatory markers, such as C-reactive protein (CRP) and interleukins, can be indicative of ongoing inflammation, including that associated with pancreatitis. Monitoring these biomarkers provides insights into the inflammatory status of the pancreas.
- *Genetic Biomarkers:* In cases of congenital lipase deficiencies or hereditary lipid disorders, genetic testing can identify specific genetic mutations that contribute to these conditions. Genetic biomarkers offer valuable information for both diagnosis and risk assessment.

The integration of lipase level measurement, imaging techniques, and biomarkers enhances the precision and depth of diagnostic assessments. These diagnostic tools collectively contribute to the early detection, accurate characterization, and effective management of various diseases related to lipase dysregulation or dysfunction.

Therapeutic Implications

Understanding the roles of lipases in health and disease opens avenues for therapeutic interventions aimed at restoring balance and mitigating the impact of lipase-related disorders. This section explores therapeutic implications, including enzyme replacement therapy, lipase inhibitors, and potential future developments.

Enzyme Replacement Therapy

Enzyme replacement therapy (ERT) is a cornerstone in managing conditions associated with lipase deficiency, particularly pancreatic insufficiency. Individuals with pancreatic disorders, such as chronic pancreatitis or cystic fibrosis, may be prescribed pancreatic enzyme supplements. These supplements, containing lipases along with other digestive enzymes, aid in the digestion and absorption of nutrients (Brennan and Saif, 2019). ERT aims to optimize nutrient absorption, prevent malnutrition, and alleviate symptoms associated with impaired lipase function. By providing exogenous lipases, ERT compensates for the insufficient endogenous enzyme production, improving the digestion of dietary fats.

Lipase Inhibitors

Lipase inhibitors represent another therapeutic approach, particularly in the context of managing obesity (Lio et al., 2020). These medications work by inhibiting the activity of pancreatic lipase, thereby reducing the absorption of dietary fats. Orlistat is an example of a lipase inhibitor commonly used for weight management. Lipase inhibitors are prescribed as part of a comprehensive weight reduction strategy, which includes dietary modifications and lifestyle changes. By limiting the absorption of dietary fats, these inhibitors contribute to calorie restriction and weight loss.

Future Developments

Precision Therapies

As our understanding of lipase biology and genetics advances, the potential for precision therapies increases. Tailoring interventions based on individual genetic profiles and the specific characteristics of lipase-related disorders may lead to more effective and personalized treatment strategies.

Targeting Lipase Dysregulation

Ongoing research focuses on identifying molecules and compounds that can modulate lipase activity, addressing dysregulation in metabolic disorders. Small molecules or biologics targeting specific aspects of lipase function may offer novel therapeutic approaches.

Gene Therapy

Gene therapy holds promise for addressing genetic lipase deficiencies. By introducing functional genes or correcting mutations associated with lipase disorders, gene therapy aims to restore normal lipase activity. This area of research may revolutionize the treatment of congenital lipase deficiencies.

Advancements in Lipase Inhibition

Ongoing efforts are directed towards refining lipase inhibitors, aiming for greater efficacy with fewer side effects. Innovations in drug development may lead to the discovery of new, more selective lipase inhibitors for managing conditions related to lipid metabolism.

Thus, the therapeutic interventions related to lipases encompass a spectrum of approaches, from enzyme replacement therapy for deficiencies to lipase inhibitors for metabolic conditions. As research progresses, the field holds exciting prospects for precision therapies and novel developments that can transform the landscape of lipase-related therapeutics.

Conclusion

Lipases and their Role in Health and Disease is exploited through the molecular intricacies of lipase biochemistry, the diverse types of lipases, and their multifaceted roles in digestion, metabolism, and immune modulation has uncovered key insights into their pivotal functions. From nutrient absorption to cellular energy balance, lipases emerge as central players in maintaining physiological homeostasis. The examination of lipases in health and their contributions to cellular function, immune modulation, and nutrient absorption underscores the critical nature of these enzymes in supporting overall well-being. The diagnostic applications, including lipase level measurement and advanced imaging techniques, provide essential tools for identifying and understanding lipase-related disorders, ranging from lipase deficiency to pancreatic diseases.

Therapeutic implications, such as enzyme replacement therapy and lipase inhibitors, highlight the practical applications of our knowledge in managing lipase-associated conditions. From addressing pancreatic insufficiency to targeting obesity, these interventions exemplify the translational potential of lipase research in clinical settings. Looking ahead, potential future developments offer exciting prospects for precision therapies and innovative treatments. Gene therapy, advancements in lipase inhibition, and a deeper understanding of the genetic basis of lipase-related disorders present avenues for tailoring interventions to individual needs, paving the way for more effective and personalized treatments. In summary, our exploration of lipases has traversed the molecular foundations to the practical applications, encompassing health, disease, diagnostics, and therapeutics. The implications of this research extend beyond the laboratory, impacting the clinic and offering hope for more targeted and personalized approaches to lipase-related conditions. As we conclude this study, the imperative for future research becomes evident, beckoning us to delve deeper into the complexities of lipase biology, uncover novel therapeutic strategies, and advance our understanding of the intricate roles these enzymes play in the intricate tapestry of human health.

References

Althaher, A. R. (2022). An Overview of Hormone-Sensitive Lipase (HSL). *The Scientific World Journal, 2022*: 1964684, 1-9. doi: 10.1155/2022/1964684.

Ater, D., & Koch, D. D. (2021). A Review of Acute Pancreatitis. *Journal of the American Medical Association, 325(23):* 2402. doi: 10.1001/jama.2021.6012.

Balasubramanian, S., Aggarwal, P., & Sharma, S. (2023). Lipoprotein Lipase Deficiency. In: *StatPearls* [Internet]. Treasure Island (FL): StatPearls Publishing; 2023 Jan-. Available from: https://www.ncbi.nlm.nih.gov/books/NBK560795/.

Bauer, E., Jakob, S., & Mosenthin, R. (2005). Principles of Physiology of Lipid Digestion. *Asian-Australasian Journal of Animal Sciences 18(2):* 282-295.

Blaak, E. E. (2003). Fatty Acid Metabolism in Obesity and Type 2 Diabetes Mellitus. *Proceedings of the Nutrition Society, 62:* 753–760. doi: 10.1079/PNS2003290.

Brennan, G. T., & Saif, M. W. (2019). Pancreatic Enzyme Replacement Therapy: A Concise Review. *Journal of Pancreas, 20(5):*121-125.

Brownlee, I., Forster, D., Wilcox, M., Dettmar, P., Seal, C., & Pearson, J. (2010). Physiological parameters governing the action of pancreatic lipase. *Nutrition Research Reviews, 23(1):* 146-154. doi: 10.1017/S0954422410000028.

Carrière, F., Thirstrup, K., Hjorth, S., & Boel, E. (1994). Cloning of the Classical Guinea Pig Pancreatic Lipase and Comparison with the Lipase Related Protein 2. *FEBS Letters, 338:* 63–68. doi: 10.1016/0014-5793(94)80117-7.

Cerk, I. K., Wechselberger, L., & Oberer, M. (2018). Adipose Triglyceride Lipase Regulation: An Overview. *Current Protein and Peptide Science, 19(2):* 221-233. doi: 10.2174/1389203718666170918160110.

Chandra, P., Enespa, Singh, R., & Arora P. K. (2020). Microbial Lipases and Their Industrial Applications: A Comprehensive Review. *Microbial Cell Factories, 19:* 169. doi: 10.1186/s12934-020-01428-8.

Cheng, H. M., Mah, K. K., & Seluakumaran, K. (2020). Fat Digestion: Bile Salt, Emulsification, Micelles, Lipases, Chylomicrons. In: Defining Physiology: Principles, Themes, Concepts Vol. 2. Springer, Cham. doi: 10.1007/978-3-030-62285-5_18.

De Koster, J., Nelli, R. K., Strieder-Barboza, C. de Souza, J., Lock A. L., Contreras, G. A. (2018). The Contribution of Hormone Sensitive Lipase to Adipose Tissue Lipolysis and Its Regulation by Insulin in Periparturient Dairy Cows. *Scientific Reports, 8:* 13378. doi: 10.1038/s41598-018-31582-4.

De Oliveira, C., Khatua, B., Noel, P., Kostenko, S., Bag, A., Balakrishnan, B., Patel, K. S., Guerra, A. A., Martinez, M. N., Trivedi, S., McCullough, A., Lam-Himlin, D. M., Navina, S., Faigel, D. O., Fukami, N., Pannala, R., Phillips, A. E., Papachristou, G. I., Kershaw, E. E., Lowe, M. E., & Singh, V. P. (2020). Pancreatic Triglyceride Lipase Mediates Lipotoxic Systemic Inflammation. *The Journal of Clinical Investigation, 130(4):*1931-1947. doi: 10.1172/JCI132767.

Flaherty, S. E., Grijalva, A., Xu, X., Ables, E., Nomani, A., & Ferrante, A. W. Jr. (2019). A lipase-independent pathway of lipid release and immune modulation by adipocytes. *Science, 1:* 363(6430):989-993. doi: 10.1126/science.aaw2586.

Garrett, R. L., & Young, R. J. (1975). Effect of Micelle Formation on the Absorption of Neutral Fat and Fatty Acids by the Chicken. *The Journal of Nutrition, 105(7):* 827-838. doi: 10.1093/jn/105.7.827.

Ghishan, F. K., Moran, J. R., Durie, P. R., & Greene, H. L. (1984). Isolated Congenital Lipase—Colipase Deficiency. *Gastroenterology, 86(6):*1580-1582. doi: 10.1016/S0016-5085(84)80175-4.

Grabner, G. F., Xie, H., Schweiger, M., & Zechner, R. (2021). Lipolysis: Cellular Mechanisms for Lipid Mobilization from Fat Stores. *Nature Metabolism, 3:* 1445–1465. doi: 10.1038/s42255-021-00493-6.

Habib, A., Chokr, D., Wan, J., Hegde, P., Mabire, M., Siebert, M., Ribeiro-Parenti, L., Le Gall, M., Lettéron, P., Pilard, N., Mansouri, A., Brouillet, A., Tardelli, M., Weiss, E., Le Faouder, P., Guillou, H., Cravatt, B. F., Moreau, R., Trauner, M., & Lotersztajn, S. (2019). Inhibition of Monoacylglycerol Lipase, an Anti-inflammatory and Antifibrogenic Strategy in the Liver. *Gut, 68(3):*522-532. doi: 10.1136/gutjnl-2018-316137.

Holm, C., Osterlund, T., Laurell, H., & Contreras, J. A. (2000). Molecular Mechanisms Regulating Hormone-Sensitive Lipase and Lipolysis. *Annual Reviews of Nutrition, 20:*365-93. doi: 10.1146/annurev.nutr.20.1.365.

Howie, D., Bokum, A. T., Necula, A. S., Cobbold, S. P., & Waldmann, H. (2017). The Role of Lipid Metabolism in T Lymphocyte Differentiation and Survival. *Frontiers in Immunology, 8*: 2017. doi: 10.3389/fimmu.2017.01949.

Hubler, M. J., & Kennedy, A. J. (2016). Role of Lipids in the Metabolism and Activation of Immune Cells. *The Journal of Nutritional Biochemistry, 34:* 1-7. doi: 10.1016/j.jnutbio.2015.11.002.

Jawed, A., Singh, G., Kohli, S., Sumera, A., Haque, S., Prasad, R., & Paul, D. (2019). Therapeutic Role of Lipases and Lipase Inhibitors Derived from Natural Resources for Remedies Against Metabolic Disorders and Lifestyle Diseases. *South African Journal of Botany, 120:* 25-32. doi: 10.1016/j.sajb.2018.04.004.

Kersten, S. (2014). Physiological Regulation of Lipoprotein Lipase. *Biochimica et Biophysica Acta, 1841(7):*919-933. doi: 10.1016/j.bbalip.2014.03.013.

Kraemer, F. B., & Shen, W. J. (2002). Hormone-Sensitive Lipase: Control of Intracellular Tri-(di-)acylglycerol and Cholesteryl Ester Hydrolysis. *Journal of Lipid Research 43:* 1585-1594. doi: 10.1194/jlr.r200009-jlr200.

Kulminskaya, N., & Oberer, M. (2020). Protein-Protein Interactions Regulate the Activity of Adipose Triglyceride Lipase in Intracellular Lipolysis. *Biochimie, 169:* 62-68. doi: 10.1016/j.biochi.2019.08.004.

Lai, H. H., Yeh, K. Y., Hsu, H. M., & Her, G. M. (2023). Deficiency of Adipose Triglyceride Lipase Induces Metabolic Syndrome and Cardiomyopathy in Zebrafish. *International Journal of Molecular Science, 24*: 117. doi: 10.3390/ijms24010117.

Li, X., & Sun, K. (2018). Regulation of Lipolysis in Adipose Tissue and Clinical Significance. *Advances in Experimental Medicine and Biology, 1090:* 199-210. doi: 10.1007/978-981-13-1286-1_11.

Li, F., & Zhang, H. (2019). Lysosomal Acid Lipase in Lipid Metabolism and Beyond. *Arteriosclerosis, Thrombosis, and Vascular Biology 39(5):* 850-856. doi: 10.1161/ATVBAHA.119.312136.

Lippi, G., Valentino, M., & Cervellin, G. (2012). Laboratory Diagnosis of Acute Pancreatitis: in Search of the Holy Grail. *Critical Reviews in Clinical Laboratory Sciences, 49(1):* 18–31. DOI: 10.3109/10408363.2012.658354.

Liu, T. T., Liu, X. T., Chen, Q. X., & Shi, Y. (2020). Lipase Inhibitors for Obesity: A Review. *Biomedicine and Pharmacotherapy 128:* 110314. doi: 10.1016/j.biopha.2020.110314.

Loli, H., Narwal, S. K., Saun, N. K., & Gupta, R. (2015). Lipases in Medicine: An Overview. *Mini-Reviews in Medicinal Chemistry, 15(14):* 1209-1216. doi: 10.2174/1389557515666150709122722.

Lotti, M., & Alberghina, L. (2007). Lipases: Molecular Structure and Function. In: Polaina, J., MacCabe, A. P. (eds) *Industrial Enzymes.* Springer, Dordrecht. DOI: 10.1007/1-4020-5377-0_16.

Meher, S., Mishra, T. S., Sasmal, P. K., Rath, S., Sharma, R., Rout, B., & Sahu, M. K. (2015). Role of Biomarkers in Diagnosis and Prognostic Evaluation of Acute Pancreatitis. *Journal of Biomarker, 2015:* 519534. doi: 10.1155/2015/519534.

Miled, N., Canaan, S., Dupuis, L., Roussel, A., Rivière, M., Carrière, F., de Caro, A., Cambillau, C., & Verger, R. (2000). Digestive Lipases: from Three-Dimensional Structure to Physiology. *Biochimie, 82(11):* 973-986. doi: 10.1016/s0300-9084(00)01179-2.

Mohy-ud-din, N., & Morrissey, S. (2023). Pancreatitis. In: *StatPearls* [Internet]. Treasure Island (FL): StatPearls Publishing; 2023 Jan-. Available from: https://www.ncbi.nlm.nih.gov/books/NBK538337/.

Monika, C., Zuzana, P., Eliska, P., Helena, D., Jana, Z., Vojtech, S., & Ludmila, K. (2012). Nutritional Regulation of Adipose Tissue Lipoprotein Lipase is Blunted in Insulin Resistant Rats. *Open Life Sciences, 7(2):* 201-209. doi: 10.2478/s11535-012-0002-y.

Nagarajan, S., Cross, E., Sanna, F., & Hodson, L. (2022). Dysregulation of Hepatic Metabolism with Obesity: Factors Influencing Glucose and Lipid Metabolism. *Proceedings of the Nutrition Society, 81(1):* 1-11. doi: 10.1017/S0029665121003761.

Pirahanchi, Y., Anoruo, M. D., & Sharma, R. (2023). Biochemistry, Lipoprotein Lipase. In: *StatPearls* [Internet]. Treasure Island (FL): StatPearls Publishing; 2023 Jan-. Available from: https://www.ncbi.nlm.nih.gov/books/NBK537040/.

Qiao, Z., Zhang, H., Zhang, Y., & Wang, K. (2020). Detection of Lipase Activity in Cells by a Fluorescent Probe Based on Formation of Self-Assembled Micelles. *iScience, 23(7):*101294. doi: 10.1016/j.isci.2020.101294.

Reddy, P., & Jialal, I. (2022). Biochemistry, Fat Soluble Vitamins. In: *StatPearls* [Internet]. Treasure Island (FL): StatPearls Publishing; 2023 Jan-. Available from: https://www.ncbi.nlm.nih.gov/books/NBK534869/.

Reynisdottir, S., Angelin, B., Langin, D., Lithell, H., Eriksson, M., Holm, C., & Arner, P. (1997). Adipose Tissue Lipoprotein Lipase and Hormone-Sensitive Lipase. Contrasting Findings in Familial Combined Hyperlipidemia and Insulin Resistance Syndrome. *Arteriosclerosis, Thrombosis, and Vascular Biology 7(10):* 2287-2292.

Sahu, A., & Birner-Gruenberger, R. (2013). Lipases. In: Kretsinger, R. H., Uversky, V. N., Permyakov, E. A. (eds) *Encyclopedia of Metalloproteins*. Springer, New York, NY. doi: 10.1007/978-1-4614-1533-6_49.

Sassene, P. J., Fanø, M., Mu, H., Rades, T., Aquistapace, S., Schmitt, B., Cruz-Hernandez, C., Wooster, T. J., & Müllertz, A. (2016). Comparison of Lipases for *In Vitro* Models of Gastric Digestion: Lipolysis Using Two Infant Formulas as Model Substrates. *Food and Function, 7:* 3989-3998. doi: 10.1039/C6FO00158K.

Subramoni, S., Suárez-Moreno, Z. R., & Venturi, V. (2010). Lipases as Pathogenicity Factors of Plant Pathogens. In: Timmis, K. N. (eds) *Handbook of Hydrocarbon and Lipid Microbiology*. Springer, Berlin, Heidelberg. doi: 10.1007/978-3-540-77587-4_248.

Tomasik, P. J., Wędrychowicz, A., Rogatko, I., Zając, A., Fyderek, K., & Sztefko, K. (2013). Gastric Lipase Secretion in Children with Gastritis. *Nutrients, 5(8):* 2924-32. doi: 10.3390/nu5082924.

Vargas, E., Joy, N. V., & Carrillo Sepulveda, M. A. (2023). Biochemistry, Insulin Metabolic Effects. In: *StatPearls* [Internet]. Treasure Island (FL): StatPearls Publishing; 2023 Jan-. Available from: https://www.ncbi.nlm.nih.gov/books/NBK525983/.

Wang, H., & Eckel, R. H. (2009). Lipoprotein Lipase: From Gene to Obesity. *American Journal of Physiology, Endocrinology and Metabolism 297:* E271–E288. DOI: 10.1152/ajpendo.90920.2008.

Zambon, A., Deeb, S.S., Pauletto, P., Crepaldi, G., & Brunzell, J. D. (2003). Hepatic Lipase: A Marker for Cardiovascular Disease Risk and Response to Therapy. *Current Opinion in Lipidology, 14(2):* 179-189.

Zechner, R., Kienesberger, P. C., Haemmerle, G., Zimmermann, R., & Lass, A. (2009). Adipose Triglyceride Lipase and the Lipolytic Catabolism of Cellular Fat Stores. *Journal of Lipid Research, 50(1):* 3-21. doi: 10.1194/jlr.R800031-JLR200.

Zechner, R., Zimmermann, R., Eichmann, T. O., Kohlwein, S. D., Haemmerle, G., Lass, A., & Madeo, F. (2012). Fat Signals-Lipases and Lipolysis in Lipid Metabolism and Signaling. *Cell Metabolism, 15(3):*279-91. doi: 10.1016/j.cmet.2011.12.018.

Zhu, G., Fang, Q., Zhu, F., Huang, D., & Yang, C. (2021). Structure and Function of Pancreatic Lipase-Related Protein 2 and Its Relationship with Pathological States. *Frontiers in Genetics, 12:* 693538. doi: 10.3389/fgene.2021.693538.

Chapter 2

Lipase: Beyond Digestion - Exploring Health Benefits, Metabolic Influence, and Therapeutic Potential

Sandhya Mulchandani[1,*] and Pratik Kale[2]

[1]Microbiology Department, Smt. Chandibai Himathmal Mansukhani College, Ulhasnagar, India
[2]K. J. Somaiya College, Mumbai, India

Abstract

Lipase is an enzyme that plays a crucial role in the digestion and absorption of dietary fats. It is produced by the pancreas and secreted into the small intestine, which breaks down triglycerides into fatty acids and glycerol, which the body can absorb. Lipase also has several health benefits beyond its digestive function. Studies have shown that lipase supplementation may improve lipid metabolism, which can lead to a reduction in body fat and an improvement in overall cardiovascular health. In addition, lipase has been shown to have anti-inflammatory properties, which may help reduce the risk of chronic diseases such as diabetes, arthritis, and certain types of cancer. Lipase may also play a role in weight management. Lipase supplementation may help reduce appetite and increase satiety, decreasing caloric intake and potentially aiding in weight loss efforts. Furthermore, lipase has been studied for its potential use in treating pancreatic insufficiency, a condition in which the pancreas cannot produce enough enzymes to digest food properly. Lipase supplementation can help alleviate abdominal pain, diarrhea, and

---

[*] Corresponding Author's Email: bharti.mul@gmail.com.

In: Lipases and their Role in Health and Disease
Editors: Vasudeo Zambare and Mohd. Fadhil Md. Din
ISBN: 979-8-89113-628-1
© 2024 Nova Science Publishers, Inc.

malabsorption symptoms. Lipase is an essential enzyme with numerous health benefits beyond its role in digestion. Supplementation with lipase may improve lipid metabolism, reduce inflammation, aid in weight management, and alleviate symptoms of pancreatic insufficiency. Further research is needed to fully understand the potential health benefits of lipase and its optimal use in clinical settings.

Keywords: lipase, health benefits, pancreas, cardiovascular health, weight management

Introduction

From the beginning, the field of catalysis, especially in relation to biological molecules like enzymes, has fascinated scholars and the industry. Historically, scientists have unearthed numerous enzymes and analysed their catalytic characteristics to elucidate intricate biological processes and manage or harness them directly. In contrast to solid catalysts, such as oxides, sulphides, halides, and metals, enzymatic catalysis offers the capacity for more intricate chemical reactions due to the precise interactions that most enzymes establish with substrates, resulting in the production of specific products (Park and Park, 2022). Enzymes are bio-catalysts found in all living things. The accuracy with which enzymes function, utilizing minimal energy to catalyse a specific process, distinguishes their biological activity. Enzymes are essential at all levels of metabolism and biological processes. Certain enzymes are intriguing due to their ability to serve as catalysts in diverse biochemical processes, offering many practical applications. They exhibit both sustainability and safety, and their specificity significantly enhances the overall efficiency of these processes (Ali et al., 2023). Hydrolytic enzymes such as proteases, amylases, amidases, esterases, and lipases constitute a substantial portion of the industrial enzyme sector. Lipases, in particular, have recently gained prominence in the rapidly expanding field of biotechnology, thanks to their multifaceted attributes that find utility across a broad spectrum of industrial sectors, including the chemical industry, biomedical sciences, food technology, and detergents (Gupta et al., 2004).

Lipase and Its Function in Digestion

Lipase also referred to as triacylglycerol acyl hydrolase, constitutes an enzyme family specializing in the breakdown of triglycerides into free fatty acids and glycerol. These enzymes are categorised within the hydrolase family and act upon carboxylic ester bonds. They are classified as serine hydrolases and do not necessitate cofactors for their function. Natural lipases carry out the hydrolysis of triglycerides, yielding fatty acids, glycerol, monoglycerides, and diglycerides. (Jaeger and Reetz, 1998; Chandra et al., 2020). Specifically, hepatic lipases are localized in the liver, while hormone-sensitive lipases are situated in adipocytes. Lipoprotein lipase is positioned on the surface of the arterial endothelium, and pancreatic lipase is present in the small intestine. The lipases found in pancreatic secretions play a pivotal role in the digestion of fats and the hydrolysis of fat-soluble vitamins for absorption. To comprehend the pathogenesis of fat necrosis and both acute and chronic pancreatitis, a comprehensive understanding of lipase function is indispensable. Moreover, lipases play a crucial role in the regulation of cholesterol-lowering medications (Pirahanchi and Sharma, 2023).

The primary phase of triglyceride and phospholipid digestion occurs in the mouth upon their interaction with saliva. The physical process of chewing, combined with the action of emulsifiers, facilitates the activity of digestive enzymes. The digestive process commences with the enzyme lingual lipase, accompanied by a small quantity of phospholipids serving as emulsifiers. These actions render the fats more accessible for enzymatic digestion, causing them to separate from the aqueous components and form minuscule droplets. Gastric lipase further breaks down triglycerides into diglycerides and fatty acids in the stomach. Approximately 30% of triglycerides undergo this conversion to diglycerides and fatty acids within a span of 2-4 hours after a meal. The stomach's churning and contractions play a role in dispersing the fat molecules, and the diglycerides produced in this process serve as additional emulsifying agents. However, it's important to note that the stomach's capacity to digest fats is relatively limited (Figure 1).

Upon the entry of stomach contents into the small intestine, a substantial portion of dietary fats remains undigested, forming large clusters. Bile, produced by the liver and stored in the gallbladder, is discharged into the duodenum, the initial segment of the small intestine. Bile salts exhibit both hydrophobic and hydrophilic properties, enabling them to attract both fats and water. This unique quality makes bile salts effective emulsifiers, breaking down large fat globules into smaller droplets. Emulsification increases the surface area available for enzymatic activity, thereby improving the accessibility of lipids to digestive enzymes. The pancreas releases pancreatic lipases into the small intestine to initiate the enzymatic digestion of triglycerides. In this process, triglycerides are enzymatically broken down into fatty acids, monoglycerides (comprising a glycerol backbone with one fatty acid still attached), and free glycerol. It is important to note that cholesterol and fat-soluble vitamins are not subject to enzymatic digestion (Callahan et al., 2020).

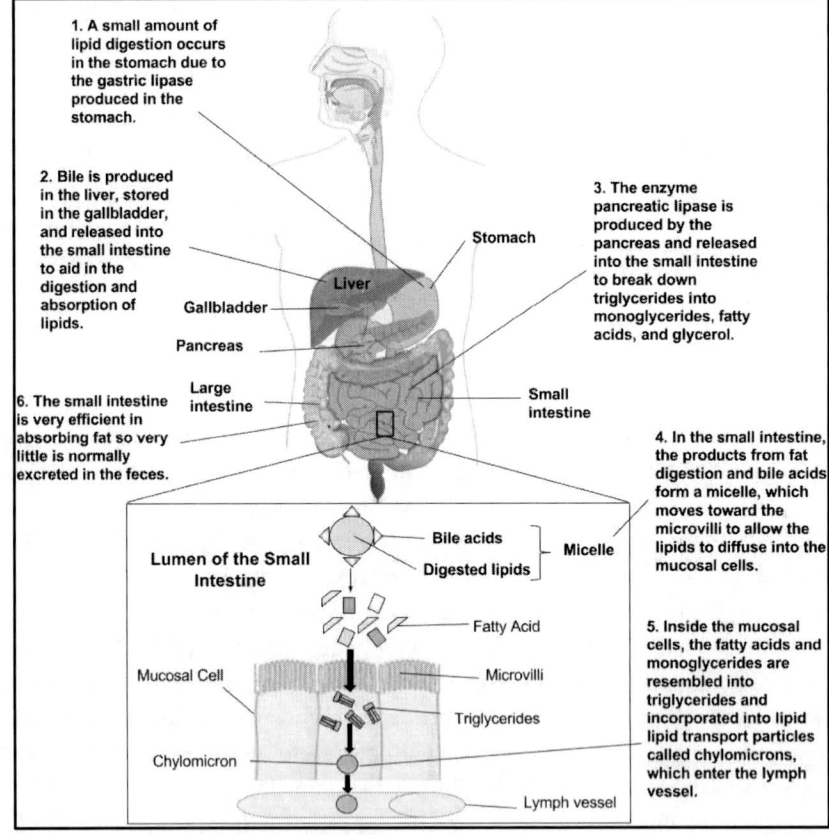

Figure 1. Lipid digestion in the human gastrointestinal tract (Titchenal et al, 2022).

Following the digestion of fats, including glycerol, fatty acids, cholesterol, fat-soluble vitamins, and monoglycerides, these byproducts need to enter the circulatory system for utilization by cells throughout the body. Once again, bile plays an integral role in this process. It encapsulates the products of fat digestion, creating micelles that assist in positioning fats in close proximity to the microvilli of intestinal cells, thereby facilitating absorption. The byproducts of fat digestion diffuse across the intestinal cell membrane, while bile salts undergo recycling to sustain their function of emulsifying fats and forming micelles (Figure 2).

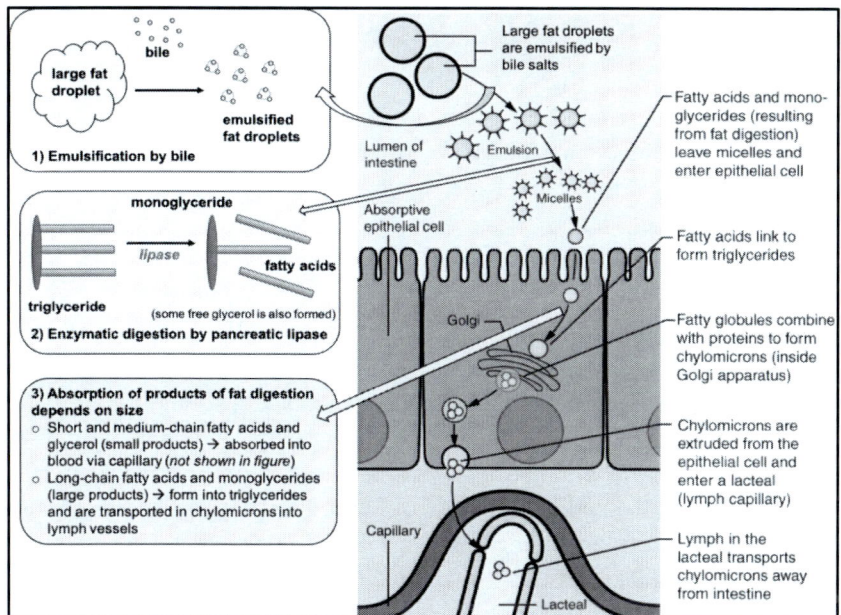

Figure 2. Lipid digestion and absorption in the small intestine (Callahan et al, 2020).

In the intestinal cell, glycerol, shorter-chain and medium-chain fatty acids, have the ability to directly enter the bloodstream. However, the absorption and transportation of larger lipids, including cholesterol, long-chain fatty acids, fat-soluble vitamins and monoglycerides, necessitate assistance. Within the intestinal cell monoglycerides and long-chain fatty acids are reconstructed into triglycerides. Together with fat-soluble vitamins and cholesterol, they become integral components of transport vehicles known as chylomicrons. These chylomicrons are substantial structures characterized by a core of cholesterol and triglycerides, enclosed by an outer membrane consisting of phospholipids interspersed with proteins known as apolipoproteins and cholesterol. The outer membrane confers water solubility to chylomicrons, facilitating their transport within the aqueous environment of the body. Chylomicrons originating from the small intestine first traverse lymph vessels before ultimately entering the bloodstream, delivering their cargo (Callahan et al, 2020; Titchenal et al., 2022).

Health Benefits of Lipases Beyond Digestion

Lipases comprise an enzyme group responsible for breaking down fats in the body, facilitating the digestion and absorption of dietary fats. Yet, their significance goes beyond digestion, as lipases have been identified as key players in various physiological processes, offering potential health advantages in realms such as metabolic disorders, immune function, and cardiovascular well-being.

Role of Lipases in Cardiovascular Diseases (CVD)

CVD stands as a significant global cause of both illness and mortality. An important factor in the development of CVD is the dysregulation of lipids and lipoproteins. The enzyme lipase plays a vital role in lipid metabolism and is closely linked to the pathophysiology of CVD. Promisingly, lipases obtained from natural sources, including medicinal plants, demonstrate the potential to prevent and manage CVD by reducing lipid levels and thwarting atherosclerosis in animal models (Jawed et al., 2019).

To comprehend the role of lipids and lipoproteins in CVD, their classification is essential. Each category plays a distinct role in the initiation and progression of CVD. High-density lipoprotein (HDL), often referred to as "good cholesterol," aids in eliminating excess cholesterol from the bloodstream. Conversely, low-density lipoprotein (LDL), known as "bad cholesterol," contributes to the formation of arterial plaque. Lipases sourced from natural origins, such as plants and microorganisms, exhibit promise in addressing metabolic disorders and lifestyle-related diseases, including CVD (Bhargava et al., 2022).

Role of Lipases in Metabolic Disorders

Metabolic disorders, including obesity, dyslipidemia, and diabetes, have emerged as substantial global health concerns. Lipases play a pivotal role in lipid metabolism, and their malfunction can contribute to the development of various metabolic disorders. Inhibitors targeting diacylglycerol lipases (DAGLs) have been identified as potential therapeutic agents for conditions like neurodegenerative and metabolic disorders. DAGLs are central to the regulation of endocannabinoid signalling, which plays roles in numerous physiological and pathological processes such as pain, inflammation, appetite, and energy metabolism. By inhibiting DAGL activity, it is possible to lower endocannabinoid levels, resulting in reduced food intake, increased energy expenditure, and enhanced glucose metabolism (Janssen and Stelt, 2016). Monoacylglycerol lipase (MAGL) inhibitors have also been recognized as potential modulators of lipid metabolism in conditions like neurological disorders, metabolic disorders, and cancer malignancy. MAGL is a crucial enzyme involved in the hydrolysis of the endocannabinoid 2-arachidonoylglycerol (2-AG), which plays a role in regulating energy balance and glucose metabolism. Inhibiting MAGL activity can elevate 2-AG levels, leading to improved glucose tolerance and insulin sensitivity. Furthermore, MAGL inhibitors have demonstrated the ability to reduce adiposity and enhance energy expenditure in animal models of obesity (Deng and Li, 2020).

Endothelial lipase (EL), another vital member of the lipase family, contributes to lipid metabolism. Predominantly expressed in vascular endothelial cells, EL regulates the metabolism of HDL and LDL cholesterol. EL can break down HDL particles, releasing free cholesterol for peripheral tissues to absorb. Conversely, EL aids in clearing LDL particles by facilitating their uptake by the liver. Dysregulation of EL activity has been linked to the development of atherosclerosis and cardiovascular disease (Yu et al., 2018). Lipase inhibitors have been developed for obesity treatment, with Orlistat being an approved pancreatic lipase inhibitor. Orlistat reduces dietary fat absorption by inhibiting pancreatic lipase activity, resulting in weight loss. Other lipase inhibitors like cetilistat and tesofensine are in development for the management of type 2 diabetes mellitus and obesity (Liu et al., 2020). In summary, lipases and their inhibitors hold significant promise in addressing metabolic disorders.

Role of Lipases in Immune Function

The Lipase enzyme plays a pivotal role in immune function. Specifically, Adipose triglyceride lipase (ATGL) is an enzyme that participates in lipid metabolism and is additionally implicated in immune response, atherosclerosis, and inflammation. ATGL deficiency in macrophages impairs their phagocytic activity and promotes the accumulation of cholesterol, leading to atherosclerosis. In contrast, ATGL overexpression in macrophages promotes cholesterol efflux and decreases inflammation. Moreover, ATGL-derived fatty acids serve as signalling molecules in the immune response and play a role in T cell activation and cytokine production (Radovic et al, 2012). Lipid metabolism also plays a crucial role in the tumour microenvironment. On one hand, lipid metabolism provides energy for immune cells and promotes their survival and proliferation. On the other hand, lipids can also promote immunosuppression and tumour growth. In particular, lipids can inhibit the function and activation of T cells and natural killer cells (Yu et al., 2021).

Lysosomal acid lipase (LAL) is an enzyme located in lysosomes also called as suicidal bag that plays a role in the breakdown of lipids. LAL deficiency leads to the accumulation of cholesterol and triglycerides in various organs and tissues, resulting in a spectrum of diseases collectively known as lysosomal storage disorders. LAL also plays a role in immune function. LAL-derived lipids can activate macrophages and promote the clearance of pathogens. Moreover, LAL deficiency in macrophages impairs their phagocytic activity and promotes inflammation (Gomaraschi et al., 2019). It can be a promising immune-metabolic target for the treatment of infectious and inflammatory diseases. Bacterial lipolysis is a mechanism used by certain bacteria to evade innate immune defences. Bacteria produce lipases that can hydrolyse host lipids, releasing free fatty acids that inhibit the function of immune cells. Moreover, bacterial lipases can also cleave immune-activating ligands, rendering them inactive and preventing their recognition by the immune system (Chen and Alonzo, 2019). Bacterial lipases are potential targets for the development of novel antimicrobial agents. Lipase enzymes play an important role in immune function and offer promising targets for the treatment of immune-mediated disorders and infectious diseases.

Importance Understanding and Potential Health Benefits of Lipases

Lipase is a multifunctional enzyme present in a variety of bodily fluids and tissues, including adipose tissues, blood, gastric juices, pancreatic secretions, and intestinal juices. Its primary function is the hydrolysis of fats, particularly triglycerides, into their component glycerol and fatty acid molecules. Triglycerides are a key source of energy for the body, and their proper metabolism is essential for overall health. However, high levels of triglycerides have been associated with an increased risk of heart disease, highlighting the critical role of lipase in the maintenance of optimal health. Lipases are involved in the regulation of lipid metabolism and can act as potential targets for developing new drugs to treat metabolic disorders. Saponins, for example, have been found to have potential health benefits in the treatment of obesity by regulating lipid metabolism through the modulation of lipase activity. Furthermore, lipid metabolism plays a significant role in the pathogenesis of various diseases, and understanding the underlying mechanisms can lead to the development of new drugs that can prevent or treat lipid-related disorders (Marrelli et al., 2016).

Exploring the potential health benefits of lipases is essential for developing effective therapeutic strategies for various diseases. Research in this area can lead to the development of new drugs that target lipid metabolism and address the growing problem of metabolic disorders such as obesity, hyperlipidemia, and atherosclerosis.

Lipases in Lipid Metabolism

Lipid metabolism is the series of biochemical reactions responsible for emulsifying the majority of dietary fats within the body into smaller particles, aided by bile. These smaller particles are subsequently broken down into free fatty acids and monoglycerides by lipases released from both the pancreas and the small intestine. In the realm of lipid metabolism, the liver and pancreas serve as vital hubs, playing pivotal roles in the processes of lipid digestion, absorption, synthesis, degradation, and transport.

Lipid Metabolism and Its Health Benefits

Lipid metabolism encompasses both the breakdown (lipolysis) and synthesis (lipogenesis) of lipids. Lipases, such as enzymes, catalyse lipolysis, breaking down lipids into fatty acids and glycerol. These components are then transported to various body tissues and organs, serving as an energy source or stored as triglycerides in adipose tissue. Lipogenesis, on the other hand, primarily occurs in the liver and adipose tissue, involving the conversion of carbohydrates into fatty acids, which are esterified to form triglycerides (Gyamfi et al, 2019). Dysregulation of lipid metabolism has been associated with the development of various diseases, including obesity, type 2 diabetes, cardiovascular diseases, and cancer. Excessive lipid accumulation in adipose tissue and other organs can lead to lipotoxicity, triggering inflammation and oxidative stress, ultimately resulting in insulin resistance and other metabolic disorders. Conversely, a healthy lipid metabolism can enhance insulin sensitivity, reduce inflammation, and mitigate the risk of chronic diseases (Amadi et al., 2022).

The role of lipid metabolism in ageing and lifespan regulation has also been a subject of study. As individuals age, changes occur in lipid metabolism, potentially leading to lipid buildup in cells and tissues. Such accumulation can lead to cellular senescence and contribute to the development of age-related conditions like cardiovascular disease, neurodegenerative diseases, and cancer. Regulating lipid metabolism, especially through methods like caloric restriction and exercise, has been shown to promote healthy ageing and extend lifespan by mitigating lipid accumulation. Caloric restriction, which involves reducing calorie intake without malnutrition, has been demonstrated to support healthy ageing and lifespan extension in various organisms, including mice and primates. This effect is likely attributable to its impact on lipid metabolism, as caloric restriction can reduce lipid accumulation in cells and tissues. Exercise has also been found to promote healthy ageing and extend lifespan. Studies indicate that regular exercise can enhance lipid metabolism by reducing lipid buildup in adipose tissue and skeletal muscle. Moreover, changes in lipid metabolism may contribute to the development of age-related diseases, such as Alzheimer's disease. Understanding the role of lipid metabolism in ageing and age-related diseases is crucial for developing interventions that encourage healthy ageing and prevent such conditions (Johnson and Stolzing, 2019). Lipoprotein Lipase (LPL) is a vital enzyme in lipid metabolism that plays a crucial role in regulating plasma lipid levels. Its primary expression occurs in adipose tissue, skeletal muscle, and the heart, where it catalyses the hydrolysis of triglycerides (TG) in circulating lipoproteins. This process generates free fatty acids (FFA) that are subsequently taken up by cells for energy utilization or storage. Dysregulation of LPL activity has been linked to various metabolic disorders, including hypertriglyceridemia, obesity, and atherosclerosis. In individuals with hypertriglyceridemia, LPL activity is diminished, resulting in an accumulation of TG-rich lipoproteins in the bloodstream. Similarly, individuals with obesity experience reduced LPL activity in adipose tissue, leading to adipose tissue accumulation and contributing to insulin resistance. Given the critical role of LPL in lipid metabolism and its association with metabolic and inflammatory disorders, there is growing interest in developing therapies that modulate LPL activity. Strategies aimed at enhancing LPL activity could be beneficial in treating hypertriglyceridemia, whereas inhibitors of LPL activity may prove useful in managing obesity and related disorders (Olivecrona, 2016).

Lipid Metabolism and Lipase Supplementation

Lipase supplementation therapy has emerged as a potential therapeutic approach for various metabolic disorders characterised by impaired lipid metabolism, such as pancreatic exocrine insufficiency, cystic fibrosis, and obesity. The primary goal of this therapy is to improve lipid digestion and absorption, thereby alleviating symptoms and improving overall health outcomes. The standards for Lipase supplementation therapy include the selection of appropriate enzyme preparations, dosing regimens, and monitoring of clinical outcomes (Layer and Keller, 2003). One emerging strategy for targeting Lipase in the management of metabolic and cardiovascular diseases is the use of exogenous Lipase supplementation. Lipase supplements derived from microbial and animal sources have been shown to improve lipid metabolism and reduce circulating triglyceride levels in both animal models and human clinical trials. These supplements may also improve insulin sensitivity and reduce inflammation, suggesting potential benefits for the management of metabolic disorders. Additionally, Lipase inhibitors derived from natural sources have also shown promise as potential therapeutics for metabolic and cardiovascular diseases. Natural compounds such as polyphenols and flavonoids have been shown to inhibit Lipase activity and improve lipid metabolism in animal models and human studies. These compounds may also have additional cardioprotective effects, such as reducing oxidative stress and inflammation, which could further contribute to their therapeutic potential (Geldenhuys et al., 2017).

The effects of Lipase supplementation were studied on subjective ratings of fullness and satiety in healthy individuals before and after consuming a high-fat meal. Participants who received Lipase supplementation reported feeling less full and less satisfied after the meal compared to those who did not receive the supplementation. Studies suggest that Lipase supplementation may interfere with normal postprandial signals of fullness and satiety, which could lead to overeating and weight gain. These findings have important implications for the use of Lipase supplementation as a therapy for improving lipid metabolism. While Lipase supplementation may have beneficial effects on lipid metabolism, it is important to consider the potential negative effects on satiety and overall food intake (Levine et al., 2015). Some studies have reported limited or inconsistent benefits, and others have raised concerns about potential adverse effects, such as gastrointestinal discomfort and impaired absorption of fat-soluble vitamins. Therefore, the effectiveness of Lipase supplementation therapy may be influenced by various factors, such as the presence of bile acids, the composition of the diet, and individual variations in enzyme activity.

Lipid Metabolism and Lipase Mechanism

There exist various types of Lipases, each with distinct functions. One such Lipase is LAL, which has been revealed to possess a unique pathophysiological role in lipid metabolism. LAL serves as a crucial enzyme within lysosomes, responsible for the hydrolysis of cholesteryl ester (CE) and TG molecules. This hydrolytic action leads to the release of free cholesterol (FC) and fatty acids (FA). The breakdown of CE and the subsequent transport of FC for storage as lipid droplets (LD) contribute to the formation of foam cells. Consequently, the liberated FC and FA from CE and TG support membrane assembly and energy production. The lipolytic products originating from LAL play a role in activating macrophages toward an M2 phenotype through processes such as fatty acid oxidation, the generation of lipid mediators, and the secretion of VLDL by hepatocytes. In both hepatocytes and macrophage foam cells, LAL is also involved in degrading lipid droplets through an autophagic process. LAL demonstrates the ability to hydrolyse aggregates of LDL in extracellular, acidic, and lytic compartments, independently of endocytic mechanisms. Moreover, catalytically active LAL is released by human macrophages. The processes involve the ATP-binding cassette transporter A1 and the endoplasmic reticulum. The LAL enzyme is encoded by the LIPA gene, and its deficiency can lead to rare lysosomal disorders such as Wolman disease and Cholesteryl Ester Storage Disease (CESD). Notably, enzyme replacement therapy for LAL has shown significant benefits in treating these disorders. LAL-mediated lysosomal lipolysis has been shown to exert influence over various facets of metabolism. For instance, it has been observed to regulate lipid mediator production, extracellular degradation of aggregated-LDL, *VLDL*, lysosomal function and autophagy, macrophage M2 polarization and adipose tissue lipolysis. (Zhang, 2019).

Lipolysis, the breakdown of triacylglycerols stored in cellular lipid droplets, plays a vital role in lipid metabolism. Recent advancements in lipolysis have revealed essential enzymes, regulatory proteins, and mechanisms that significantly impact cellular metabolism (Grabner et al., 2021). The products and intermediates of lipolysis actively participate in cellular signalling processes, particularly in non-adipose tissues, highlighting the significance of "lipolytic signalling" that was previously underestimated. This newly recognized aspect of lipolysis holds potential relevance for human diseases (Zechner et al., 2012). Both canonical and noncanonical enzymatic pathways facilitate intracellular fatty acid mobilization. These pathways are intricately regulated by various factors and signalling pathways at both transcriptional and post-transcriptional levels. Intracellular lipolysis has been found to impact energy homeostasis, metabolic regulation, and lipid-mediated signalling across multiple organs. Moreover, these processes can contribute to the pathogenesis of diseases, making them potential targets for prevention or treatment. For instance, targeting lipolysis could be a viable strategy for addressing obesity and insulin resistance (Zhang, 2019; Grabner et al., 2021).

Lipases in Cardiovascular Health

Lipids, including cholesterol and triglycerides, are essential components of cell membranes and are involved in various physiological functions such as energy storage, hormone synthesis, and cell signalling. However, abnormal lipid metabolism can lead to the accumulation of lipids in the bloodstream and the development of cardiovascular diseases such as atherosclerosis, coronary artery disease, and myocardial infarction. Therefore, maintaining optimal lipid metabolism through the regulation of Lipase activity is an important therapeutic strategy for preventing and managing cardiovascular diseases.

Cardiovascular Disease and Its Risk Factors

CVD encompasses a spectrum of conditions that affect both the heart and blood vessels. This includes conditions like cerebrovascular disease, congenital heart disease, deep vein thrombosis, peripheral arterial disease, pulmonary embolism, coronary heart disease, and rheumatic heart disease. The leading behavioural risk factors for heart disease and stroke are unhealthy habits, including, a sedentary lifestyle, excessive alcohol consumption, an inadequate diet and tobacco use. These behaviours can lead to conditions such as high blood glucose levels, overweight, obesity, elevated blood pressure, and increased blood lipid levels., collectively known as "intermediate risk factors." Monitoring these factors in primary care settings can help identify an elevated risk of heart attack, stroke, heart failure, and other complications. In addition to behavioural risk factors, there are other factors that contribute to the risk of CVD. These include factors such as age, gender, genetic predisposition, and a family history of cardiovascular disease. Having a first-degree relative who developed CVD at a young age is considered a significant risk factor. Environmental factors, such as exposure to air pollution and poor sleep patterns, have also been linked to an increased risk of CVD.

Cardiovascular Health and Lipase Effect

Lipase, specifically hepatic Lipase (HL), is an enzyme synthesized by the liver that plays a crucial role in modifying the composition and metabolism of lipoproteins present in the bloodstream. These lipoproteins encompass VLDL, LDL, and HDL. The primary role of HL is to facilitate the clearance of triglycerides carried by VLDL, which, when elevated in the bloodstream, can increase the risk of coronary heart disease (CHD). HL accomplishes this by breaking down triglycerides within VLDL and converting them into LDL. However, LDL transports cholesterol from the liver to other tissues, and excessive LDL levels can contribute to the formation of arterial plaques, thereby elevating the risk of CHD. HL also influences the composition of LDL by removing phospholipids, resulting in smaller and denser particles that may be more susceptible to oxidation and inflammation. In contrast, HDL transports cholesterol from tissues back to the liver for disposal and is associated with a reduced risk of CHD. Nevertheless, HL modifies HDL by removing phospholipids and triglycerides, leading to smaller HDL particles with a reduced capacity to transport cholesterol. Intriguingly, some studies suggest that HL may enhance HDL's ability to protect against oxidation and inflammation. Several factors can affect HL activity and function. Genetic variations or mutations in the LIPC gene, responsible for encoding HL, can influence the production or effectiveness of HL. Additionally, hormones (such as insulin, thyroid hormones, and estrogen), dietary choices, medications (including alcohol, niacin, and statins), as well as conditions like diabetes, obesity, and inflammation can impact HL activity. Given its involvement in lipoprotein metabolism and its significance as a marker for cardiovascular disease risk and therapy response, HL has substantial implications for cardiovascular health (Zambon et al., 2003).

In people with hypertriglyceridemia, high HL activity leads to the formation of small, dense LDL particles (sdLDL), which are more likely to cause atherosclerosis, or the buildup of plaque in the arteries. In these individuals, reducing HL activity with drugs can reduce sdLDL and lower coronary artery disease (CAD) risk. In contrast, in people with hypercholesterolemia, high HL activity does not increase CAD risk because their LDL particles are large and buoyant, which are less harmful to the arteries. Increasing HL activity with drugs in these individuals can enhance the reverse cholesterol transport (RCT), or the removal of excess cholesterol from the tissues by HDL. This can lower CAD risk. In people with normal triglyceride and cholesterol levels but with the sdLDL phenotype, high HL activity can increase CAD risk by reducing HDL levels and impairing RCT. Genetic variants that affect HL activity can also influence CAD risk. One common variant is a single nucleotide polymorphism (SNP) in the promoter region of the HL gene, which controls how much HL is produced. The T allele of this SNP reduces HL promoter activity and lowers HL levels in the blood. The association between this allele and higher CAD risk varies among different population studies and may depend on the interaction between the SNP and the lipoprotein phenotype. The impact of hepatic Lipase on CAD risk varies depending on the individual's lipoprotein phenotype and genetic factors. The effect of high HL activity on CAD can be either beneficial or harmful, depending on whether it leads to the formation of small, dense LDL particles or promotes reverse cholesterol transport (Brunzell et al, 2012).

LPL is a key enzyme that breaks down lipoproteins and releases fatty acids (FA) for the heart to use as energy. While FA is the primary energy source for the heart, it can also have negative effects on cardiovascular health, especially in individuals with diabetes. Diabetes impairs the body's ability to use glucose, another energy source, due to insufficient or ineffective insulin. When glucose metabolism is disrupted, the heart becomes more reliant on FA for energy production. However, this increased dependence on FA can cause several problems for the heart, including interfering with glucose uptake and oxidation, increasing oxidative stress and inflammation, and altering the composition and structure of cardiac membranes (Lee et al., 2023). To cope with the increased demand for FA during diabetes, the heart upregulates its LPL activity by post-translational mechanisms such as phosphorylation and glycosylation. However, this may also exacerbate the harmful effects of FA on cardiac metabolism and function. LPL can generate lysophospholipids from lipoproteins, which are signalling molecules that can activate various receptors and pathways involved in cell growth, survival, differentiation, and inflammation. Some of these effects may be beneficial for the heart, such as promoting angiogenesis and cardioprotection, but others may be detrimental, such as inducing hypertrophy and apoptosis. Therefore, LPL is an enzyme that affects lipoprotein metabolism and cardiovascular disease risk. While LPL is crucial for delivering FA to the heart for energy production, its overexpression can exacerbate the harmful effects of FA on cardiac metabolism and function. LPL gene variants can affect plasma lipid levels and cardiovascular disease risk, but their effects are modulated by environmental factors such as dietary fat intake. Future studies should consider both genetic and environmental factors when investigating the role of LPL in cardiovascular disease (Lee et al., 2023; Pulinilkunnil and Rodrigues, 2006).

Cardiovascular Health and Lipase Mechanisms

Lipase can modulate the metabolism and transport of lipids and lipoproteins, which are complex particles that carry cholesterol, triglycerides, phospholipids, and apolipoproteins in the blood. Lipids and lipoproteins have diverse effects on CVDs, depending on their type, size, composition, and function. One of the studied lipoproteins in relation to CVDs is low-density lipoprotein (LDL), which is also known as "bad" cholesterol. LDL can deposit in the arterial wall and form atherosclerotic plaques, which can narrow or block the blood flow and cause ischemic events, such as myocardial infarction (MI) or stroke. Lipase can influence the level and composition of LDL by regulating its synthesis, secretion, uptake, and degradation. Oxidized LDL can trigger inflammatory responses and promote plaque formation and instability. Therefore, Lipase can enhance the atherogenic potential of LDL and increase the risk of CVDs. Another lipoprotein that is involved in CVDs is HDL, which is also known as "good" cholesterol. HDL can protect against CVDs by facilitating RCT, which is the process of removing excess cholesterol from peripheral tissues and delivering it to the liver for excretion. HDL can also exert anti-oxidant, anti-thrombotic, anti-inflammatory, anti-apoptotic, and vasodilatory effects on various cells and tissues involved in CVDs. Lipase can modulate the quantity and quality of HDL by influencing its biogenesis, maturation, remodelling, and functionality. For example, lipoprotein Lipase can increase the size and cholesterol content of HDL by transferring triglycerides from VLDL to HDL; however, it can also impair the RCT capacity of HDL by reducing its ability to accept cholesterol from macrophages (Soppert et al., 2020).

A third lipoprotein that is associated with CVDs is lipoprotein (a) [Lp(a)], which is a variant of LDL that contains an additional protein called apolipoprotein (a). Lp(a) has a dual role in CVDs: it can promote atherosclerosis by competing with plasminogen for binding to fibrin and impairing fibrinolysis, which is the process of dissolving blood clots; it can also induce calcification of the aortic valve by activating osteoblast-like cells and increasing calcium deposition. Lipase can affect the level and function of Lp(a) by modulating its production, clearance, and interaction with other molecules. For example, endothelial Lipase can reduce the plasma concentration of Lp(a) by enhancing its catabolism; however, it can also increase the prothrombotic activity of Lp(a) by removing phospholipids from its surface and exposing apo(a) (Nordestgaard and Langsted, 2016). Understanding the mechanism of Lipase's effect on cardiovascular health can help to identify novel biomarkers and therapeutic targets for CVDs. Further studies are needed to elucidate the complex interactions between Lipase and other factors that influence lipid homeostasis and CVD outcomes.

Lipases in Anti-Inflammation

Lipase has demonstrated its ability to influence inflammatory responses in various tissues and cells across the body. Through the regulation of lipid metabolism and signalling molecules derived from lipids, Lipase can effectively exert anti-inflammatory actions, playing a role in resolving inflammation.

Inflammation and Its Role in Chronic Disease

Inflammation is a natural and beneficial response of the immune system designed to safeguard the body from harmful agents such as pathogens, toxins, or damaged cells. Nevertheless, when inflammation becomes persistent and chronic, it can inflict severe damage on healthy tissues and organs, contributing to the onset of various diseases (Figure 3) (Furman et al., 2019). Chronic inflammation is linked to a multitude of diseases that affect diverse organs and systems within the body, including diabetes, obesity, cardiovascular disease, neurodegenerative diseases, autoimmune disorders, and cancer. The persistence of chronic inflammation can stem from prolonged exposure to inflammatory triggers, disturbances in immune cell function or inflammatory mediators, genetic or epigenetic factors, and environmental or lifestyle elements. Chronic inflammation can lead to tissue damage, fibrosis, angiogenesis, metabolic dysfunction, and alterations in gene expression, resulting in pathological transformations and the loss of function in affected organs (Fleit, 2014). A critical player in chronic inflammation is the macrophage, a key cellular component capable of adopting various phenotypes and functions depending on the surrounding microenvironment and signals they receive. Macrophages are broadly categorized into two types: M1 and M2. M1 macrophages are pro-inflammatory, producing cytokines like tumour necrosis factor-α (TNF-α), interleukin-1β (IL-1β), and interleukin-6 (IL-6), as well as generating reactive oxygen species and nitric oxide. They play a role in pathogen elimination, debris clearance, and initiation of adaptive immunity. In contrast, M2 macrophages are anti-inflammatory, secreting cytokines such as interleukin-10 (IL-10) and transforming growth factor-β (TGF-β), along with growth factors and extracellular matrix proteins. M2 macrophages are involved in tissue repair, wound healing, and immunoregulation. Maintaining the equilibrium between M1 and M2 macrophages is essential for preserving tissue homeostasis. However, disruptions in this balance due to chronic exposure to inflammatory stimuli or irregularities in immune cells or inflammatory mediators can lead to the development of chronic inflammation, giving rise to a variety of chronic inflammatory diseases affecting various organs and systems within the body (Fleit, 2014; Chen et al., 2017).

Anti-Inflammatory Properties of Lipases

Lipase has been discovered to possess anti-inflammatory characteristics. Inhibiting MAGL, a specific type of Lipase has emerged as a potential approach to mitigate inflammation and fibrosis in the liver. Research has demonstrated that MAGL inhibition can effectively reduce inflammation and fibrosis in the livers of mice afflicted with non-alcoholic steatohepatitis, a condition characterised by liver inflammation and damage. Consequently, targeting MAGL holds promise as a potential treatment strategy for liver diseases associated with inflammation and fibrosis (Habib et al., 2018).

Additionally, apart from MAGL inhibition, peptides with Lipase inhibitory properties, generated through gastrointestinal breakdown of heat-treated millet grains, have displayed potential anti-inflammatory effects (Jakubczyk et al., 2019). These peptides have been shown to inhibit Lipase activity, which is implicated in the production of pro-inflammatory molecules. This study suggests that the consumption of millet grains may confer anti-inflammatory benefits due to the presence of lipase-inhibitory peptides.

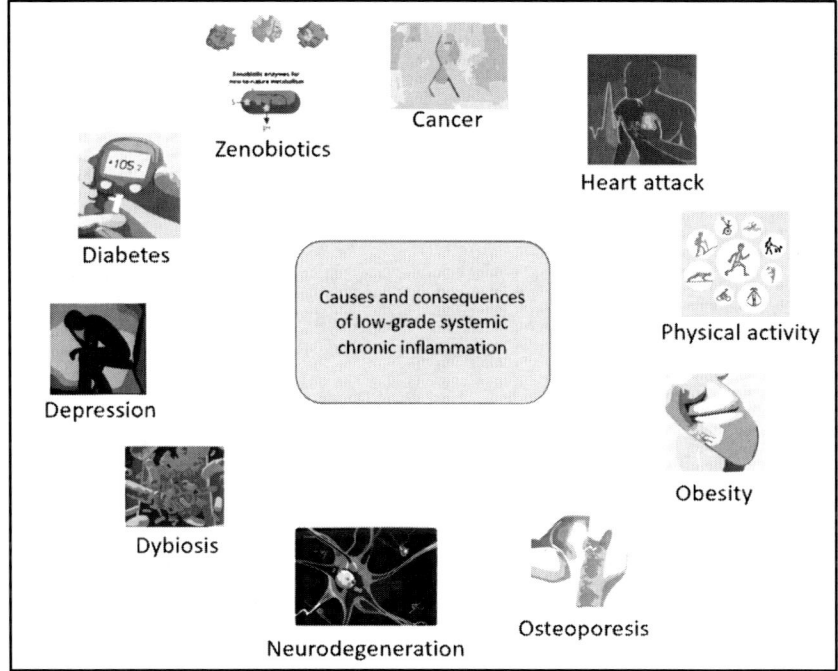

Figure 3. Causes and consequences of low-grade systemic chronic inflammation.

Moreover, the specific elimination of Endothelial Lipase (EL), another type of lipase, has been observed to increase the levels of high-density lipoprotein (HDL) particles characterized by anti-inflammatory properties. HDL is acknowledged for its anti-inflammatory attributes, and enhancing its levels could prove beneficial in alleviating inflammation. Mice deficient in EL showed elevated levels of HDL particles with anti-inflammatory characteristics compared to wild-type mice. These results suggest that targeting EL could potentially serve as a strategy to enhance HDL particles with anti-inflammatory properties (Hara et al., 2011). In summary, inhibiting lipase activity, particularly through the inhibition of MAGL, has been identified as a potential approach to reduce inflammation across various conditions. Additionally, Lipase inhibitory peptides derived from millet grains and the targeted manipulation of EL have shown promise in increasing the presence of anti-inflammatory HDL particles.

Anti-Inflammatory Effects and Lipase Mechanisms

Lipase exhibits several anti-inflammatory mechanisms, one of which involves the inhibition of MAGL. MAGL is an enzyme responsible for breaking down endocannabinoids, lipid mediators known for their anti-inflammatory and analgesic effects through the activation of cannabinoid receptors. Lipase, by inhibiting MAGL, elevates endocannabinoid levels while reducing the production of pro-inflammatory cytokines like TNF-α and IL-6 in the liver. This action contributes to the mitigation of liver fibrosis, a chronic condition characterised by excessive extracellular matrix deposition and liver tissue scarring (Habib et al., 2018). Another anti-inflammatory mechanism of Lipase involves the generation of peptides with anti-inflammatory and Lipase inhibitory properties during gastrointestinal hydrolysis of heat-treated millet grains. Millet, a cereal crop, contains proteins with high nutritional and bioactive potential. Heat treatment alters the protein structure, enhancing digestibility and peptide release during gastrointestinal hydrolysis. These peptides inhibit enzymes involved in inflammation, such as COX-1, COX-2, and LOX, responsible for synthesizing inflammatory mediators like prostaglandins and leukotrienes. Additionally, these peptides inhibit pancreatic Lipase, reducing dietary fat absorption and serum cholesterol and triglyceride levels, which are cardiovascular disease risk factors (Jakubczyk et al., 2019).

A third anti-inflammatory mechanism involves the modulation of HDL particles by endothelial Lipase (EL). EL hydrolyses phospholipids and triglycerides in HDL, which possesses anti-inflammatory and anti-oxidative properties safeguarding the endothelium from oxidative stress and inflammation. Deleting the EL gene in mice enhances HDL particle size and number, imparting additional anti-inflammatory properties both in vitro and in vivo. EL-deficient HDL promotes endothelial nitric oxide production, inhibits vascular cell adhesion molecule-1 expression, reduces monocyte adhesion to endothelial cells, and decreases atherosclerotic lesion development in mice (Hara et al., 2011). A fourth anti-inflammatory mechanism involves the analgesic and anti-inflammatory effects of MAGL inhibition observed in mice with collagen-induced arthritis (CIA), a rheumatoid arthritis model. RA, a chronic inflammatory joint disorder, causes pain, swelling, stiffness, and deformities. MAGL inhibition using the selective inhibitor JZL184 elevates endocannabinoid levels, activating cannabinoid receptors in the spinal cord and peripheral tissues, thereby reducing mechanical hypersensitivity and thermal hyperalgesia induced by CIA in mice. Additionally, MAGL inhibition suppresses pro-inflammatory cytokine production in the serum and joint tissues of CIA mice, attenuating joint inflammation and cartilage damage (Nass, 2015).

A fifth potential mechanism through which Lipase inhibitors may exert anti-inflammatory effects involves the modulation of lipid mediator levels, including prostaglandins, leukotrienes, and resolvins. These molecules, derived from fatty acids, play pivotal roles in inflammation regulation and resolution. Lipase inhibitors may influence the balance of these mediators by affecting their biosynthesis or degradation, thereby impacting their biological functions. Another possible mechanism is the modulation of gut microbiota, the microorganism community residing in the digestive tract. The gut microbiota can impact inflammation by producing metabolites, modulating immune cells, and interacting with the intestinal barrier. Lipase inhibitors may alter the gut microbiota's composition and function by modifying the availability of dietary fats, a primary energy source for these microorganisms. This alteration may have implications for inflammatory bowel diseases such as Crohn's disease and ulcerative colitis (Liu et al., 2020). Lipase possesses anti-inflammatory properties that modulate the inflammatory response and prevent tissue damage in various diseases.

Lipases in Weight Management

In recent years, researchers have explored the potential of Lipase as a tool for weight management. Specifically, some studies have suggested that Lipase supplements may help individuals lose weight by enhancing fat metabolism and reducing fat absorption.

Weight Management and Its Health Benefits

The management of weight entails the process of attaining and sustaining a healthy weight level to diminish the risk of developing chronic illnesses and enhance overall quality of life. Effective weight management includes the adoption of a well-balanced diet, regular engagement in physical activity, and consistent monitoring of weight and body composition. This approach to weight management is crucial for maintaining good health because an excess of body weight, particularly in cases of obesity, is linked to heightened morbidity and mortality across various conditions, including cardiovascular disease, diabetes, cancer, gallstones, osteoarthritis, sleep apnea, and respiratory issues (Doucet et al., 2021; Aphramor, 2010). According to the American Heart Association, maintaining a healthy weight has several benefits for the circulatory system, such as improving blood flow, regulating fluid levels, and preventing diabetes and heart disease. Moreover, maintaining a healthy weight can also enhance one's self-esteem, energy levels, and motivation to make other positive health changes. Body Mass Index (BMI) serves as a valuable screening tool to evaluate an individual's weight status and associated health risks. Derived by dividing an individual's weight in kilograms by the square of their height in meters, BMI offers a numerical measure. A BMI falling within the range of 18.5 to 24.9 is categorized as normal, whereas a BMI between 25 and 29.9 denotes overweight status. Additionally, a BMI of 30 or higher signifies obesity, indicating an elevated risk of health complications.

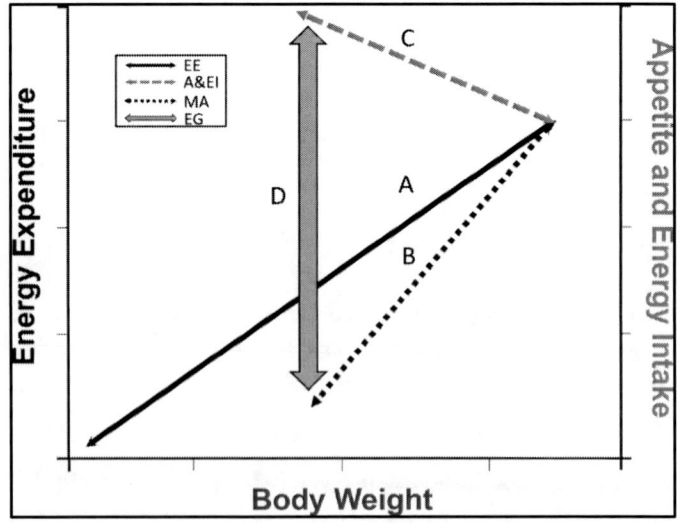

Figure 4. Illustration depicting the impact of weight loss on different factors, including energy expenditure (EE) (A), metabolic adaptations (MA) (referring to a more substantial decrease in EE than predicted) (B), appetite and energy intake (A&EI) (C), and the energy gap (EG) (D). (Doucet et al., 2021).

Energy balance is a crucial factor in the context of weight control, as it revolves around the interplay between the calories obtained from food and drinks (energy intake) and the calories expended through metabolic processes and physical activity (energy expenditure). As demonstrated in Figure 4, the decline in energy expenditure (A), coupled with metabolic adjustments (B), alongside an increased urge to eat (C), affects the ability to uphold energy balance after losing weight. It's plausible that weight loss progressively exacerbates the energy gap (D), making it increasingly difficult to sustain equilibrium as each additional kilogram is shed (Doucet et al., 2021). To maintain a healthy weight, there must be a balance between energy intake and expenditure. Weight loss necessitates that energy intake is lower than expenditure, while weight gain requires the opposite. Energy balance is influenced by various factors, including genetics, environment, behaviour, and metabolism. Achieving a healthy weight involves adhering to a nutritious eating regimen that provides essential nutrients while restricting excess calories, fats, sugars, and salt. This regimen should encompass a variety of foods from different food groups, such as whole grains, lean protein sources, nuts, seeds, fruits, vegetables, low-fat dairy products, and oils. Moreover, it should be tailored to individual needs, preferences, and objectives. Physical activity is another crucial aspect of weight management, capable of elevating energy expenditure, enhancing cardiovascular fitness, fortifying muscles and bones, reducing stress and depression, and preventing weight regain. Additionally, physical activity can have favourable impacts on blood cholesterol levels, blood pressure, insulin sensitivity, and blood sugar levels. Effective weight management can aid in the prevention and treatment of numerous chronic ailments while enhancing overall well-being. By adhering to a healthful eating plan, engaging in regular physical activity, and closely monitoring weight and body composition, individuals can attain and uphold a healthy weight conducive to heart health.

Weight Management and Lipase Effect

Lipase inhibitors refer to substances that diminish the activity or expression of Lipase, thereby reducing the absorption of fat and enhancing its elimination from the body. These inhibitors have garnered attention as potential agents for combating obesity, as they hold the promise of assisting in weight reduction and enhancing metabolic parameters. Among the most widely recognized Lipase inhibitors is orlistat, a synthetic pharmaceutical that operates by inhibiting gastric and pancreatic Lipases within the gastrointestinal tract. The molecular structure of orlistat is depicted in Figure 5. Clinical trials lasting up to two years have demonstrated that orlistat is more effective than dietary modifications alone in achieving weight loss and sustaining it. Additionally, orlistat treatment has been associated with modest enhancements in parameters such as low-density lipoprotein levels, total cholesterol, fasting glucose and insulin levels, as well as blood pressure. Nonetheless, orlistat is not without its limitations, which encompass gastrointestinal side effects, diminished absorption of fat-soluble vitamins, and a moderate degree of weight loss. Notably, common adverse effects of orlistat include issues like faecal urgency, incontinence, flatus with discharge, and oily spotting. Furthermore, orlistat has the potential to interfere with the absorption of certain medications, such as cyclosporine, levothyroxine and warfarin (Heck et al., 2000).

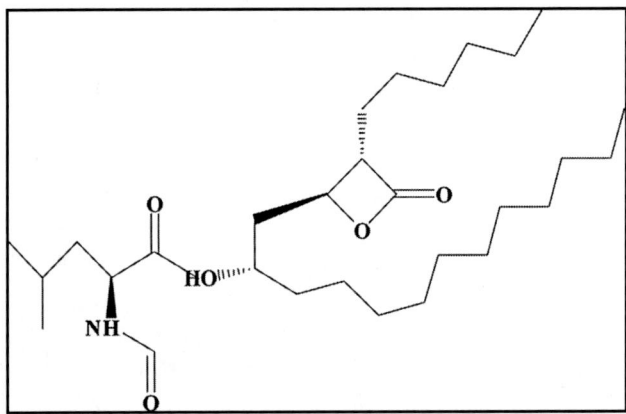

Figure 5. The molecular structure of orlistat (Liu et al., 2020).

Natural products have emerged as promising reservoirs of Lipase inhibitors, primarily owing to their structural diversity, low toxicity, and widespread accessibility. In a review conducted by Liu et al., in 2020, a comprehensive exploration of various natural Lipase inhibitors derived from sources like plants, fungi, bacteria, and animals was undertaken. The review delved into their mechanisms of action, pharmacological impacts, and safety profiles. Among the array of natural Lipase inhibitors, examples include green tea catechins, saponins, flavonoids, terpenoids, alkaloids, and peptides. For instance, green tea catechins have exhibited the ability to inhibit pancreatic Lipase activity in vitro to curtail fat absorption and mitigate body weight gain in mice subjected to a high-fat diet. Similarly, saponins extracted from diverse plants have showcased Lipase inhibitory prowess in vitro and manifested anti-obesity effects in vivo. Cinnamon, a commonly used spice in both culinary and traditional medicinal contexts, represents another natural product explored for its lipase-inhibitory potential. In a study conducted by Khedr et al., (2019), the effects of orlistat and cinnamon on weight management and metabolic parameters in obese rats were compared. The findings demonstrated that both orlistat and cinnamon resulted in reduced body weight gain, diminished food intake, lowered serum lipid levels, decreased liver enzyme levels, and a decline in oxidative stress markers among obese rats. Notably, cinnamon supplementation also led to elevated serum adiponectin levels, a hormone known for enhancing insulin sensitivity and fostering fatty acid oxidation. This observation suggested that cinnamon might offer supplementary benefits for weight management by modulating adipokine secretion and mitigating inflammation. Lipase inhibitors encompass both synthetic and natural varieties, each with its unique advantages and disadvantages. Nonetheless, further research is imperative to assess the long-term safety and efficacy of Lipase inhibitors in the context of weight management, as well as their potential interactions with other medications or nutrients.

Weight Management and Lipase Mechanism

Obesity is a prevalent and pressing health issue in contemporary society, often stemming from unhealthy lifestyles characterised by excessive fat consumption and a lack of physical activity. The accumulation of surplus fats within the body primarily impacts the liver and white adipose tissue, which are responsible for lipid storage. This excess lipid buildup can lead to the enlargement of non-alcoholic fatty liver and white adipose tissue. Obesity and hyperlipidemia contribute to a range of risk factors, including hypertension, glucose intolerance, insulin resistance, and liver fibrosis, all of which heighten the risk of mortality. Research conducted on individuals with type 2 diabetes has suggested that the accumulation of lipids within muscle cells precedes hepatic insulin resistance and the onset of diabetes. The development of obesity is intricately tied to the metabolism of body fat, with a significant portion of the diet consisting of mixed triglycerides. However, the human body cannot directly utilize exogenous fat and necessitates hydrolysis for absorption. Within the digestive system, several Lipases, including those in the tongue, stomach, and pancreas, play crucial roles. Pancreatic Lipase, in particular, is indispensable as it performs a pivotal function in hydrolysing dietary lipids, converting triacylglycerol substrates into monoglycerides and free fatty acids within the digestive system. These monoglycerides and free fatty acids are subsequently absorbed by enterocytes, the cells lining the intestines.

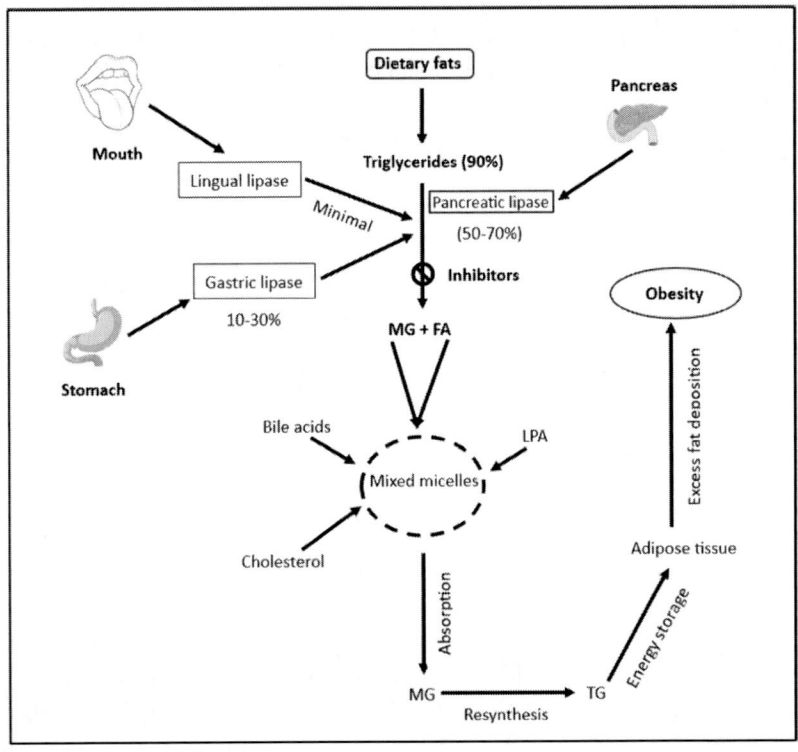

Figure 6. Lipid metabolic pathways in the human body (Adapted from Tsagkaris et al., 2022).

After the consumption of fat-containing food, Lipase comes into play by breaking down triglycerides into monoglycerides, glyceryl esters, and free fatty acids. While lingual Lipase contributes minimally to fat degradation, it plays a significant role in breaking down 50% to 70% of fat intake in infants and young children. Gastric Lipase (accounting for 10% to 30% of decomposition) and pancreatic Lipase (responsible for 50% to 70% of decomposition) further hydrolyse fats within the gastrointestinal tract and small intestine. This process leads to the formation of cholesterol, lipoproteins, and lipid mixed particles. These components, along with bile acids, are subsequently absorbed by the small intestine, and any excess energy is stored as adipose tissue through the re-synthesis of triacylglycerol. Figure 6 provides an overview of the lipid metabolic pathway within the human body. Lipase inhibitors play a vital role in this context by binding to the active Lipase component within the stomach and small intestine. This binding leads to conformational changes in stomach/trypsin, resulting in the inhibition of catalytic activity. Consequently, the hydrolysis and digestion of lipids, such as triglycerides, are reduced, thereby decreasing lipid absorption from food and preventing the accumulation of adipose tissue. Notably, Lipase inhibitors are typically excreted along with the Lipase they've bound to, causing no long-term effects on the human body. Currently, weight loss medications primarily fall into two categories: peripheral Lipase inhibitors and central appetite suppressants. Peripheral Lipase inhibitors, like orlistat, operate by inhibiting the intestinal absorption of fat. In contrast, central appetite suppressants, including fenfluramine and sibutramine, act on the central nervous system. However, clinical studies have identified adverse reactions associated with these medications, encompassing headaches, dizziness, dry mouth, bitterness, constipation, and insomnia. Additionally, these medications may lead to various mental or cardiovascular adverse reactions, resulting in limitations in their clinical applications and, in some cases, withdrawal from the market. Peripheral Lipase inhibitors offer a relatively safe alternative, as they do not enter the bloodstream or affect the nervous system, mineral balance, or bone circulation. Consequently, Lipase inhibitors have been considered a safer option for development (Liu et al., 2020).

Lipases in Pancreatic Insufficiency

Pancreatic insufficiency is when the pancreas does not produce enough Lipase or other digestive enzymes. This can lead to malabsorption, weight loss, and other symptoms. Pancreatic insufficiency can be caused by various diseases or disorders that affect the pancreas, such as cystic fibrosis, chronic pancreatitis, or pancreatic cancer.

Pancreatic Insufficiency and Its Symptoms

Pancreatic insufficiency is a condition in which the pancreas does not produce enough digestive enzymes to break down food properly. Digestive enzymes are essential for the absorption of nutrients from food, especially fats, proteins, and vitamins (Capurso et al., 2019; Domínguez-Muñoz, 2011). When these enzymes are deficient, food is not digested well and passes through the intestines undigested or partially digested. This can cause various symptoms and complications, such as:

- Steatorrhea: This is the most common symptom of pancreatic insufficiency. It refers to the presence of excess fat in the stool, which makes it oily, bulky, foul-smelling, and difficult to flush. Steatorrhea can lead to weight loss, malnutrition, dehydration, and electrolyte imbalance.
- Abdominal pain: This can occur due to the accumulation of undigested food in the intestines, which can cause inflammation, bloating, gas, and cramps. Abdominal pain can also be caused by pancreatic inflammation or infection.
- Diarrhoea: This can result from the rapid transit of undigested food through the intestines, which can irritate the intestinal lining and cause fluid loss. diarrhoea can also be caused by bacterial overgrowth in the small intestine due to reduced acidity and bile salt levels.
- Flatulence: This can occur due to the fermentation of undigested carbohydrates by bacteria in the colon, which can produce gas and cause bloating and discomfort.
- Weight loss: This can occur due to the reduced absorption of calories and nutrients from food, as well as increased energy expenditure due to malabsorption and diarrhoea.
- Vitamin deficiencies: This can occur due to the reduced absorption of fat-soluble vitamins (A, D, E, and K) and water-soluble vitamins (B12 and folate). Vitamin deficiencies can cause various symptoms and complications, such as anaemia, bleeding disorders, night blindness, osteoporosis, and neurological problems (Capurso et al., 2019; Domínguez-Muñoz, 2011).

The diagnosis of pancreatic insufficiency is based on the clinical history, physical examination, nutritional assessment, and laboratory tests. The most commonly used laboratory test is the measurement of faecal elastase-1 (FE-1), which is a marker of pancreatic enzyme secretion. A low level of FE-1 indicates pancreatic insufficiency. Other tests that can be used to diagnose pancreatic insufficiency include faecal fat quantification, 13C-triglyceride breath test, secretin stimulation test, and endoscopic ultrasound. The treatment of pancreatic insufficiency consists of oral administration of pancreatic enzyme replacement therapy (PERT), which contains Lipase, amylase, and protease. PERT helps to digest food and improve symptoms and nutritional status. PERT should be taken with every meal and snack, and the dose should be adjusted according to the fat content of the diet and the response to therapy. PERT should be taken in the form of enteric-coated mini microspheres or micro tablets that resist gastric acid and release enzymes in the duodenum. PERT may also require supplementation with acid-suppressing drugs or bile salts to enhance its efficacy (Capurso et al., 2019; Domínguez-Muñoz, 2011). Pancreatic insufficiency is a serious condition that can impair the quality of life and increase the mortality of patients with pancreatic diseases. Early diagnosis and appropriate treatment with PERT can improve the clinical outcomes and reduce the complications associated with malabsorption and malnutrition.

Pancreatic Insufficiency and Lipase Supplementation

The standard treatment for pancreatic insufficiency (PI) is Lipase supplementation therapy (LST), which involves the consumption of oral capsules or tablets containing Lipase derived from pigs before or during meals and snacks. The primary objective of LST is to compensate for the deficient or insufficient Lipase in the body, thereby reinstating normal fat digestion and absorption. The prescribed dosage of LST is determined by various factors, including the severity of the PI, the quantity and type of dietary fat, the patient's body weight and age, and their response to the treatment. In general, it is recommended to initiate LST with 30,000-40,000 Lipase units per meal and 15,000 to 20,000 units for snacks, with adjustments made based on clinical symptoms and stool fat content. To ensure effectiveness, LST should be ingested with a glass of water or juice, and the capsules or tablets should be swallowed whole without chewing or crushing. It's important to avoid exposing these capsules or tablets to high temperatures or humidity, as it can affect their stability and efficacy. While Lipase supplementation therapy is typically well-tolerated and safe, there are potential side effects that may occur, including abdominal pain, bloating, flatulence, diarrhoea, constipation, nausea, vomiting, allergic reactions, and hyperuricemia (elevated uric acid levels in the blood). To minimize these side effects, it's advisable to commence treatment with a low dose and gradually increase it, take LST alongside food, and avoid overeating or excessive fat consumption (Table 1) (Layer and Keller, 2003).

LST is not effective for everyone with PI. Some patients may have other factors that impair fat digestion and absorption, such as low intestinal pH, reduced bile salt secretion, bacterial overgrowth, or intestinal inflammation. In these cases, LST may need to be combined with other therapies, such as proton pump inhibitors (to increase intestinal pH), bile salt supplements (to enhance micelle formation), antibiotics (to treat bacterial overgrowth), or anti-inflammatory drugs (to reduce intestinal inflammation). LST is also not the only option for treating PI. There are some alternatives and perspectives that may offer advantages over porcine Lipase. These include:

- Non-porcine Lipase: Some patients may have ethical or religious objections to using porcine Lipase or may develop allergic reactions to it. Non-porcine Lipase derived from fungal or bacterial sources may be suitable for these patients. However, these sources may have lower activity or stability than porcine Lipase or may have different substrate specificity or pH optimum.
- Enteric-coated Lipase: Some patients may experience gastric irritation or inactivation of Lipase by stomach acid when taking LST. Enteric-coated Lipase is designed to resist gastric acid and release Lipase in the duodenum where it can act on fats. However, enteric-coated Lipase may have variable release times depending on gastric emptying and intestinal transit times or may not release enough Lipase to achieve adequate fat digestion (Layer and Keller, 2003).
- Liprotamase: Liprotamase is a synthetic enzyme that contains three components: amylase (to digest carbohydrates), protease (to digest proteins), and Lipase (to digest fats). It is intended to mimic the natural pancreatic enzyme secretion and provide a complete enzyme replacement therapy for PI. However, liprotamase has not been approved by the Food and Drug Administration (FDA) due to insufficient evidence of efficacy and safety in clinical trials.
- Gene therapy: Gene therapy is a promising approach that aims to correct the underlying genetic defect that causes PI in some patients, such as those with cystic fibrosis. Gene therapy involves delivering a functional copy of the defective gene into the target cells using a vector (such as a virus). This could potentially restore normal pancreatic function and eliminate the need for LST. However, gene

therapy is still in its early stages of development and faces many challenges, such as safety issues, immune responses, gene delivery efficiency, gene expression regulation, and ethical concerns.

Table 1. PERT management (Brennan and Saif, 2019)

Changing the lifestyle	Smoking and alcohol intake must be discontinued entirely.
Dietary routine	It is advisable to follow a nutritious diet, and there is no need for strict limitations on dietary fat. Eating smaller, more frequent meals could be beneficial in reducing symptoms.
Begin treatment with an appropriate PERT dosage.	30,000 to 40,000 international units (IU) for main meals and 15,000 to 20,000 IU for snacks.
The timing is crucial.	PERT should be consumed alongside meals and snacks, and if multiple capsules are necessary, they should be ingested at various intervals prior to, during, and following meals.
Make sure to have a follow-up and evaluate the response.	Monitoring the therapeutic response is of utmost importance. In cases where patients do not show improvement, a thorough evaluation should be conducted to explore other potential causes, including pancreatic cancer (assessed via EUS and imaging), celiac disease (examined using TTG IgA), giardia (detected through stool antigen testing), SIBO (evaluated via a breath test), irritable bowel syndrome, and mucosal disease (investigated through endoscopy).
Gradual improvement of PERT in a systematic manner.	1. Evaluate adherence and stress the importance of taking the prescribed dose with every meal and snack. 2. Raise the PERT dosage (double the amount). 3. Consider a trial of concurrent PPI or H2B therapy. 4. Switch to a different formulation.

LST is an effective and well-established treatment for PI that can improve nutritional status and quality of life for many patients. However, LST is not a one-size-fits-all solution and may need to be individualized according to each patient's needs and preferences. Moreover, LST is not without limitations and drawbacks that may warrant further research and development of alternative or complementary therapies for PI (Layer and Keller, 2003; Fieker et al., 2011).

Pancreatic Insufficiency and Lipase Mechanisms

The mechanisms of Lipase's effects on pancreatic insufficiency can be understood from different perspectives: the physiological role of Lipase in fat digestion, the pharmacokinetics and pharmacodynamics of Lipase supplementation, and the clinical outcomes of Lipase supplementation therapy. Lipase is secreted by both the gastric and pancreatic glands in response to various stimuli, such as the cephalic, gastric, and intestinal phases of digestion. Gastric Lipase initiates lipolysis in the stomach by hydrolysing triglycerides into diglycerides and free fatty acids. Pancreatic Lipase continues lipolysis in the duodenum by hydrolysing triglycerides and diglycerides into monoglycerides and free fatty acids. The products of lipolysis are then emulsified by bile salts and phospholipids to form micelles, which can be transported across the unstirred water layer and taken up by the enterocytes. In patients with pancreatic insufficiency, the activity of pancreatic Lipase is reduced or absent, resulting in impaired fat digestion and absorption. Gastric Lipase may partially compensate for the pancreatic defect, but its activity is also regulated by interacting neuro-hormonal mechanisms that may be impaired in pancreatic insufficiency (Wøjdemann et al., 2000).

Lipase supplementation therapy aims to provide exogenous Lipase to replace the deficient endogenous Lipase and restore normal fat digestion and absorption. Lipase supplementation products are derived from porcine pancreas and contain a mixture of digestive enzymes, including Lipase, amylase, and protease. The products are formulated as enteric-coated microspheres or mini microspheres that are resistant to gastric acid and release the enzymes in the duodenum. The efficacy of Lipase supplementation therapy depends on several factors, such as the dose, timing, pH, mixing, transit time, and interaction with food and drugs. The optimal dose of Lipase supplementation therapy is determined by titrating the dose according to the patient's symptoms, body weight, fat intake, and degree of steatorrhea. The optimal timing of Lipase supplementation therapy is to administer the product with each meal or snack, preferably at the beginning or during the meal. The optimal pH for Lipase activity is between 6 and 8; therefore, acid suppression with proton pump inhibitors or histamine-2 receptor antagonists may enhance Lipase activity and efficacy. The optimal mixing of Lipase supplementation therapy with food is achieved by thoroughly chewing the food or using liquid formulations. The optimal transit time for Lipase supplementation therapy is between 30 and 60 minutes; therefore, prokinetic agents may improve Lipase efficacy by accelerating gastric emptying and intestinal transit. The interaction of Lipase supplementation therapy with food and drugs may affect its efficacy by altering its bioavailability or activity. For example, high-fat meals may increase the demand for Lipase activity; calcium-containing antacids may impair the dissolution of enteric coating; and bile acid sequestrants may reduce micelle formation (Layer and Keller, 2003).

The clinical outcomes of Lipase supplementation therapy are measured by various parameters, such as adverse events, pain relief, quality of life, bone mineral density, vitamin K, vitamin E, vitamin D, vitamin A, serum triglycerides, serum cholesterol, serum prealbumin, serum albumin, BMI, body weight, FE-1, fecal fat excretion, coefficient of nitrogen absorption (CNA), and fat absorption coefficient (FAC). The primary goal of Lipase supplementation therapy is to achieve an FAC of at least 90%, which indicates normal fat absorption. A secondary goal is to achieve a CNA of at least 80%, which indicates normal protein absorption. Other parameters reflect the nutritional status, metabolic status, micronutrient status, bone health status, functional status, symptom relief status, and safety profile of Lipase supplementation therapy. Several studies have shown that Lipase supplementation therapy can improve most of these parameters in patients with pancreatic insufficiency due to various causes, such as chronic pancreatitis, cystic fibrosis, pancreatic surgery, or enteral nutrition supplementation (Freedman, 2017; Fieker et al., 2011; Brennan and Saif., 2019).

Potential Side Effects of Lipase Supplementation

Lipase is an enzyme that helps break down fats in the digestive system. Some people may take Lipase supplements to improve their digestion or to treat certain medical conditions. However, Lipase supplementation may also have some potential side effects, such as abdominal pain, nausea, diarrhoea, or allergic reactions. Lipase supplements are often used to aid digestion, particularly in individuals with pancreatic disorders like cystic fibrosis and chronic pancreatitis. However, these supplements can have potential side effects that should be considered before use. One of the most significant risks of Lipase supplements is the possibility of an allergic reaction, especially if the supplement contains other enzymes such as amylase or protease. An allergic reaction can cause symptoms like hives, itching, swelling, difficulty breathing, and anaphylactic shock, which can be life-threatening. Therefore, people who are allergic to pork, beef, or any other source of Lipase should avoid taking Lipase supplements or consult their doctor before using them. Another potential side effect of Lipase supplements is neurological problems, such as headaches, dizziness, insomnia, and mood changes. These effects may be caused by the interaction of Lipase with other enzymes or substances in the supplement or in the body. For example, Lipase may affect the levels of serotonin, a neurotransmitter that regulates mood and sleep. People who have a history of neurological disorders or who are taking medications that affect the nervous system should be cautious about using Lipase supplements and monitor their symptoms carefully.

Lipase supplements may also cause gastrointestinal issues, such as nausea, diarrhea, cramping, and stomach pain. These effects may be due to the increased breakdown of fats in the digestive tract, which can irritate the lining of the stomach and intestines. Lipase supplements may also interfere with the absorption of some nutrients, such as fat-soluble vitamins (A, D, E, and K) and essential fatty acids. Therefore, people who take Lipase supplements should also take a multivitamin and omega-3 supplement to prevent deficiencies. It is important to note that Lipase supplements are not recommended for everyone and should be used with caution and under medical supervision. People who have a medical condition that affects the pancreas or digestion, such as cystic fibrosis, chronic pancreatitis, pancreatic cancer, or steatorrhea (fatty stools), should consult their doctor before taking Lipase supplements. People who are pregnant or breastfeeding should also avoid taking Lipase supplements unless prescribed by their doctor. Furthermore, Lipase supplements may interact with some medications such as blood thinners, diabetes drugs, or cholesterol-lowering drugs. Therefore, people who take these medications should also check with their doctor before using Lipase supplements (Keller, 2005).

Factors Affecting Risk or Side Effects of Lipase Supplementation

Some of the factors that increase the risk of side effects associated with lipase supplementation are:

- Allergy or sensitivity to Lipase or other ingredients in the supplement: Some people may experience allergic reactions such as hives, itching, swelling, difficulty breathing, or anaphylaxis after taking Lipase supplements. This can be life-threatening and requires immediate medical attention. To prevent this, people should check the label of the supplement and avoid it if they are allergic or sensitive to any of the ingredients.
- Interactions with other medications or supplements: Lipase supplements may interact with some medications or supplements and

affect their effectiveness or safety. For example, Lipase supplements may reduce the absorption of some fat-soluble vitamins (A, D, E, and K) and minerals (calcium, iron, magnesium, and zinc). They may also interfere with some blood thinners (such as warfarin), antidiabetic drugs (such as metformin), cholesterol-lowering drugs (such as statins), and anti-inflammatory drugs (such as ibuprofen).
- Underlying medical conditions or diseases: Lipase supplements may worsen some medical conditions or diseases that affect the digestive system or other organs. For example, Lipase supplements may increase the risk of bleeding or bruising in people with bleeding disorders or liver problems. They may also aggravate symptoms of gastritis, ulcers, irritable bowel syndrome, or inflammatory bowel disease. They may also cause hyperlipidemia (high levels of fats in the blood) or hyperuricemia (high levels of uric acid in the blood) in some people.
- Dosage and duration of use: Lipase supplements may cause some side effects if taken in higher doses than recommended or for longer periods than necessary. Some of the common side effects of Lipase supplementation are nausea, vomiting, diarrhea, abdominal pain, cramps, bloating, gas, indigestion, heartburn, and bad breath. These side effects are usually mild and temporary and can be reduced by taking Lipase supplements with meals or snacks and drinking plenty of water. However, if these side effects persist or worsen, people should stop taking Lipase supplements and seek medical help (Keller, 2005).

Lipase supplementation can be beneficial for some people with certain digestive disorders, but it can also cause some side effects if not used properly. People who want to take Lipase supplements should be aware of the factors that increase the risk of side effects and take precautions to avoid them.

Recommendation for Safe Use of Lipase Supplementation

a. Selection and Standardization of Lipase Supplements

When selecting Lipase supplements, it is crucial to consider certain factors to ensure their safety and efficacy. According to (Layer and Keller,

2003), one important aspect is the standardization of Lipase products. This involves establishing specific quality criteria such as potency, stability, and bioavailability. Standardized Lipase supplements are manufactured consistently, ensuring that each dose contains the intended amount of active enzyme. Reliable sources for Lipase supplements should be preferred, and manufacturers should follow stringent quality control measures to guarantee their therapeutic effectiveness.

 b. Optimizing Lipase Dosage and Administration

Determining the appropriate dosage of Lipase supplementation is crucial to achieving optimal clinical outcomes (Fieker et al., 2011). Individualizing Lipase dosage based on several patient-specific factors is crucial. Body weight, dietary fat intake, and the severity of malabsorption are key considerations. Healthcare professionals should calculate the Lipase dosage based on the patient's body weight, ensuring that an adequate amount of Lipase is administered to facilitate proper digestion and absorption of fats. Adjustments to the dosage may be required over time, as patients' conditions and dietary needs can change. Regular monitoring of symptoms and laboratory parameters can help assess the effectiveness of the current dosage and guide any necessary adjustments.

 c. Considerations for Enteral Nutrition Supplementation

When managing exocrine pancreatic insufficiency (EPI) in patients receiving enteral nutrition supplementation, incorporating Lipase supplementation is a vital consideration. According to Freedman's study, various options can be explored to address EPI in these individuals (Freedman, 2017). Lipase supplementation is essential to facilitate the breakdown and absorption of fats in the enteral nutrition formula. Healthcare providers should ensure that the Lipase dosage is adjusted accordingly, considering the fat content of the formula and the patient's specific nutritional requirements. Additionally, optimizing the delivery methods of enteral nutrition can enhance the effectiveness of Lipase supplementation and improve overall nutrient absorption.

d. Assessment of Pancreatic Exocrine Function

Assessing pancreatic exocrine function is crucial in determining the need for Lipase supplementation. (Keller, 2005) study suggests the use of diagnostic tests to evaluate pancreatic function accurately. Fecal elastase-1 measurement is a commonly used non-invasive test that assesses the activity of pancreatic enzymes. Low fecal elastase-1 levels indicate reduced pancreatic exocrine function and support the need for Lipase supplementation. Healthcare professionals can use these tests to diagnose exocrine pancreatic insufficiency and determine the appropriate dosage of Lipase supplementation.

e. Safety and Adverse Effects

While Lipase supplementation is generally safe, it is essential to be aware of potential adverse effects. Monitoring patients for gastrointestinal symptoms is crucial to identify any intolerances or side effects associated with Lipase supplementation. Common symptoms include abdominal pain, diarrhea, bloating, and flatulence. If these symptoms occur, healthcare providers should assess whether they are related to Lipase supplementation and consider adjusting the dosage or exploring alternative options. Continuous monitoring and close communication with patients are necessary to ensure their safety and well-being throughout the course of Lipase supplementation therapy.

Lipase supplementation therapy offers a valuable approach for managing pancreatic insufficiency and optimizing lipid digestion. Based on insights from the aforementioned studies, recommendations for the safe use of Lipase supplementation include selecting standardized Lipase products, individualizing dosage based on patient characteristics, considering enteral nutrition supplementation, assessing pancreatic exocrine function, and monitoring for potential adverse effects. These recommendations can guide healthcare professionals in effectively incorporating Lipase supplementation as part of a comprehensive treatment approach for conditions associated with impaired lipid digestion.

Conclusion

Lipase, a versatile enzyme with crucial roles in health, holds promise in several key areas of research. In the pharmaceutical field, thermostable

Lipases have diverse applications, from drug manufacturing to bioenergy. Lipase-based diagnostics provide valuable tools for detecting conditions like pancreatitis and skin disorders. LPL activators show potential in managing metabolic disorders and obesity, offering tailored solutions. Lipases may also have a role in cancer treatment. In medical devices, immobilized Lipases enhance fat absorption in patients with conditions like cystic fibrosis. In clinical settings, Lipase-based technologies can inform treatment decisions and enhance patient care. Continued research and innovation in this field promise to advance disease management, treatment outcomes, and our understanding of metabolic processes, ultimately benefiting individuals and populations.

References

Ali, S., Khan, S. A., Hamayun, M., & Lee, I. J. (2023). The recent Advances in the Utility of Microbial Lipases: A Review. *Microorganisms, 11*(2): 510. doi: 10.3390/micro organisms11020510.

Amadi, P. U., Gu, H.-M., Yin, K., Jiang, X. C., & Zhang, D-W. (2022). Editorial: Lipid Metabolism and Human Diseases. *Frontiers in Physiology, 13*. doi: 10.3389/fphys.2022.1072903.

Aphramor, L. (2010). Validity of Claims Made in Weight Management Research: A Narrative Review of Dietetic Articles. *Nutrition Journal, 9*(1): 30. doi: 10.1186/1475-2891-9-30.

Bhargava, S., de la Puente-Secades, S., Schurgers, L., & Jankowski, J. (2022). Lipids and Lipoproteins in Cardiovascular Diseases: A Classification. *Trends in Endocrinology & Metabolism, 33(6):*409-423. doi: 10.1016/j.tem.2022.02.001.

Brennan, G. T., & Saif, M. W. (2019). Pancreatic Enzyme Replacement Therapy: A Concise Review. *Journal of the Pancreas, 20*(5).

Brunzell, J. D., Zambon, A., & Deeb, S. S. (2012). The Effect of Hepatic Lipase on Coronary Artery Disease in Humans Is Influenced by the Underlying Lipoprotein Phenotype. *Biochimica et Biophysica Acta - Molecular and Cell Biology of Lipids, 1821(3):* 365-372. doi: 10.1016/j.bbalip.2011.09.008.

Callahan, A., Leonard, H., & Powell, T. (2020). Nutrition: Science and Everyday Application. *Open Oregon.* https://openoregon.pressbooks.pub/nutritionscience/.

Capurso, G., Traini, M., Piciucchi, M., Signoretti, M., Arcidiacono, PG. (2019). Exocrine Pancreatic Insufficiency: Prevalence, Diagnosis, and Management. *Clinical and Experimental Gastroenterology, 12:* 129-139. doi: 10.2147/ceg.s168266.

Chandra, P., Enespa, Singh, R., & Arora, P. K. (2020). Microbial Lipases and Their Industrial Applications: A Comprehensive Review. *Microbial Cell Factories, 19 (1).* doi: 10.1186/s12934-020-01428-8.

Chen, L., Deng, H., Cui, H., Fang, J., Zuo, Z., Deng, J., Li, Y., Wang, X., & Zhao, L. (2017). Inflammatory Responses and Inflammation-Associated Diseases in Organs. *Oncotarget, 9(6):* 7204-7218. doi: 10.18632/oncotarget.23208.

Chen, X., & Alonzo, F. (2019). Bacterial Lipolysis of Immune-Activating Ligands Promotes Evasion of Innate Defenses. *Proceedings of the National Academy of Sciences, 116(9):* 3764-3773. doi: 10.1073/pnas.1817248116.

Deng, H., & Li, W. (2020). Monoacylglycerol lipase inhibitors: Modulators for Lipid Metabolism in Cancer Malignancy, Neurological and Metabolic Disorders. *Acta Pharmaceutica Sinica B, 10(4):* 582-602. doi: 10.1016/j.apsb.2019.10.006.

Domínguez-Muñoz, J. E. (2011). Pancreatic Exocrine Insufficiency: Diagnosis and Treatment. *Journal of Gastroenterology and Hepatology, 26:* 12-16. doi: 10.1111/j.1440-1746.2010.06600.x.

Doucet, E., Hall, K., Miller, A., Taylor, V. H., Ricupero, M., Haines, J., Brauer, P., Gudzune, K. A., House, J. D., Dhaliwal, R. (2021). Emerging Insights in Weight Management and Prevention: Implications for Practice and Research. *Applied Physiology, Nutrition, and Metabolism, 46(3):* 288-293. doi: 10.1139/apnm-2020-0585.

Fieker, A., Philpott, J., & Armand, M. (2011). Enzyme Replacement Therapy for Pancreatic Insufficiency: Present and Future. *Clinical and experimental gastroenterology, 4*, 55–73. doi: 10.2147/CEG.S17634.

Fleit, H. B. (2014). Chronic Inflammation. In: *Pathobiology of Human Disease*, 300-314. doi: 10.1016/b978-0-12-386456-7.01808-6.

Freedman, S. D. (2017). Options for Addressing Exocrine Pancreatic Insufficiency in Patients Receiving Enteral Nutrition Supplementation. *AJMC.* https://www.ajmc.com/view/options-for-addressing-exocrine-pancreatic-insufficiency-in-patients-receiving-enteral-nutrition-supplementation-article.

Furman, D., Campisi, J., Verdin, E., Carrera-Bastos, P., Targ, S., Franceschi, C., Ferrucci, L., Gilroy, D. W., Fasano, A., Miller, G. W., Miller, A. H., Mantovani, A., Weyand, C. M., Barzilai, N., Goronzy, J. J., Rando, T. A., Effros, R. B., Lucia, A., Kleinstreuer, N., & Slavich, G. M. (2019). Chronic Inflammation in the Etiology of Disease across the Life Span. *Nature Medicine, 25(12):* 1822-1831. doi: 10.1038/s41591-019-0675-0.

Geldenhuys, W. J., Lin, L., Darvesh, A. S., & Sadana, P. (2017). Emerging Strategies of Targeting Lipoprotein Lipase for Metabolic and Cardiovascular Diseases. *Drug Discovery Today, 22(2):* 352-365. doi: 10.1016/j.drudis.2016.10.007.

Gomaraschi, M., Bonacina, F., & Norata, G. D. (2019). Lysosomal acid Lipase: From Cellular Lipid Handler to Immunometabolic Target. *Trends in Pharmacological Sciences, 40(2):* 104-115. doi: 10.1016/j.tips.2018.12.006.

Grabner, G. F., Xie, H., Schweiger, M., & Zechner, R. (2021). Lipolysis: Cellular Mechanisms for Lipid Mobilization from Fat Stores. *Nature Metabolism, 3(11):* 1445-1465. doi: 10.1038/s42255-021-00493-6.

Gupta, R., Gupta, N., & Rathi, P. (2004). Bacterial lipases: An Overview of Production, Purification and Biochemical Properties. *Applied Microbiology and Biotechnology, 64(6):* 763-781. doi: 10.1007/s00253-004-1568-8.

Gyamfi, D., Awuah, E. O., & Owusu, S. (2019). Lipid Metabolism. In: *The Molecular Nutrition of Fats*, 17-32. doi: 10.1016/b978-0-12-811297-7.00002-0.

Habib, A., Chokr, D., Wan, J., Hegde, P., Mabire, M., Siebert, M., Ribeiro-Parenti, L., Gall, M. L., Letteron, P., Pillard, N., Mansouri, A., Brouillet, A., Tardelli, M., Weiss, E., Faouder, PL., Guillou, H., Cravatt, B. F., Moreau, R., Trauner, M., & Lotersztajn, S. (2018). Inhibition of Monoacylglycerol Lipase, an Anti-Inflammatory and Antifibrogenic Strategy in the Liver. *Gut, 68(3):* 522-532. doi: 10.1136/gutjnl-2018-316137.

Hara, T., Ishida, T., Kojima, Y., Tanaka, H., Yasuda, T., Shinohara, M., Toh, R., Hirata, K. (2011). Targeted Deletion of Endothelial Lipase Increases HDL Particles with Anti-Inflammatory Properties Both in Vitro and in Vivo. *Journal of Lipid Research, 52(1):* 57-67. doi: 10.1194/jlr.m008417.

Heck, A. M., Yanovski, J. A., & Calis, K. A. (2000). Orlistat, a New Lipase Inhibitor for the Management of Obesity. *Pharmacotherapy, 20*(3): 270-279. doi: 10.1592/phco.20.4.270.34882.

Jaeger, K., & Reetz, M. T. (1998). Microbial Lipases form Versatile Tools for Biotechnology. *Trends in Biotechnology, 16(9):* 396-403. doi: 10.1016/s0167-7799(98)01195-0.

Jakubczyk, A., Szymanowska, U., Karaś, M., Zlotek, U., & Kowalczyk, D. (2019). Potential Anti-Inflammatory and Lipase Inhibitory Peptides Generated by in Vitro Gastrointestinal Hydrolysis of Heat-Treated Millet Grains. *CyTA - Journal of Food, 17(1):* 324-333. doi: 10.1080/19476337.2019.1580317.

Janssen, F. J., & van der Stelt, M. (2016). Inhibitors of Diacylglycerol Lipases in Neurodegenerative and Metabolic Disorders. *Bioorganic & Medicinal Chemistry Letters, 26(16):* 3831-3837. doi: 10.1016/j.bmcl.2016.06.076.

Jawed, A., Singh, G., Kohli, S., Sumera, A., Haque, S., Prasad, R., & Paul, D. (2019). Therapeutic Role of Lipases and Lipase Inhibitors Derived from Natural Resources for Remedies Against Metabolic Disorders and Lifestyle Diseases. *South African Journal of Botany, 120:* 25-32. doi: 10.1016/j.sajb.2018.04.004.

Johnson, A. A., & Stolzing, A. (2019). The Role of Lipid Metabolism in Ageing, Lifespan Regulation, and Age-related Disease. *Ageing Cell, 18(6):* e13048. doi: 10.1111/acel.13048.

Keller, J. (2005). Human Pancreatic Exocrine Response to Nutrients in Health and Disease. *Gut, 54(suppl_6):* 1-28. doi: 10.1136/gut.2005.065946.

Keller, J., & Layer, P. (2003). Pancreatic Enzyme Supplementation Therapy. *Current Treatment Options in Gastroenterology, 6(5):* 369-374. doi: 10.1007/s11938-003-0039-0.

Khedr, N. F., Ebeid, A. M., & Khalil, R. M. (2019). New Insights into Weight Management by Orlistat in Comparison with Cinnamon as a Natural Lipase Inhibitor. *Endocrine, 67(1):* 109-116. doi: 10.1007/s12020-019-02127-0.

Layer, P., & Keller, J. (2003). Lipase Supplementation Therapy: Standards, Alternatives, and Perspectives. *Pancreas, 26(1):* 1-7. doi: 10.1097/00006676-200301000-00001.

Lee, C. S., Zhai, Y., & Rodrigues, B. (2023). Changes in Lipoprotein Lipase in the Heart Following Diabetes Onset. *Engineering, 20:* 19-25. doi: 10.1016/j.eng.2022.06.013.

Levine, M. M., Koch, S. Y., & Koch, K. L. (2015). Lipase Supplementation before a High-Fat Meal Reduces Perceptions of Fullness in Healthy Subjects. *Gut and Liver, 9(4):* 464-469. DOI: 10.5009/gnl14005.

Liu, T. T., Liu, X. T., Chen, Q. X., & Shi, Y. (2020). Lipase Inhibitors for Obesity: A Review. *Biomedicine & Pharmacotherapy, 128,* 110314. doi: 10.1016/j.biopha.2020.110314.

Marrelli, M., Conforti, F., Araniti, F., & Statti, G. (2016). Effects of Saponins on Lipid Metabolism: A Review of Potential Health Benefits in the Treatment of Obesity. *Molecules, 21(10):* 1404. doi: 10.3390/molecules21101404.

Nass, S. R. (2015). Analgesic and Anti-Inflammatory Effects of Monoacylglycerol Lipase Inhibition in Mice Subjected to Collagen-Induced Arthritis (Graduate Theses, Dissertations, and Problem Report No. 6297).

Nordestgaard, B. G., & Langsted, A. (2016). Lipoprotein (a) as a Cause of Cardiovascular Disease: Insights from Epidemiology, Genetics, and Biology. *Journal of Lipid Research, 57(11):* 1953-1975. doi: 10.1194/jlr.r071233.

Olivecrona, G. (2016). Role of Lipoprotein Lipase in Lipid Metabolism. *Current Opinion in Lipidology, 27(3):* 233-241. doi: 10.1097/mol.0000000000000297.

Park, J.-Y., & Park, K.-M. (2022). Lipase and Its Unique Selectivity: A Mini-review. *Journal of Chemistry, 2022.* doi: 10.1155/2022/7609019.

Pirahanchi, Y., & Sharma, S. (2023). Biochemistry, Lipase. *StatPearls.* https://www.ncbi.nlm.nih.gov/books/NBK470341/.

Pulinilkunnil, T., & Rodrigues, B. (2006). Cardiac Lipoprotein Lipase: Metabolic Basis for Diabetic Heart Disease. *Cardiovascular Research, 69(2):* 329-340. doi: 10.1016/j.cardiores.2005.09.017.

Radovic, B., Aflaki, E., & Kratky, D. (2012). Adipose Triglyceride Lipase in Immune Response, Inflammation, and Atherosclerosis. *Biological Chemistry, 393(9):* 1005–1011. doi: 10.1515/hsz-2012-0192.

Soppert, J., Lehrke, M., Marx, N., Jankowski, J., & Noels, H. (2020). Lipoproteins and Lipids in Cardiovascular Disease: From Mechanistic Insights to Therapeutic Targeting. *Advanced Drug Delivery Reviews, 159:* 4-33. doi: 10.1016/j.addr.2020.07.019.

Titchenal, A., Hara, S., Caacbay, N. A., Meinke-Lau, W., Yang, Y-Y., Revilla, M., Draper, J., Langfelder, G., Gibby, C., Chun, C. N., Calabrese, A., Lim, E. T., & Taguchi, K. (2022). *Human Nutrition 2e.* University of Hawai'i at Mānoa Food Science and Human Nutrition Program. Simple Book Publishing. https://pressbooks.oer.hawaii.edu/humannutrition2/.

Tsagkaris, A. S., Louckova, A., Jaegerova, T., Tokarova, V., & Hajslova, J. (2022). The In Vitro Inhibitory Effect of Selected Asteraceae Plants on Pancreatic Lipase Followed by Phenolic Content Identification through Liquid Chromatography High Resolution Mass Spectrometry (LC- HRMS). *International Journal of Molecular Science, 23:* 11204. doi: 10.3390/ijms231911204.

Wøjdemann, M., Sternby, B., & Larsen, S. (2000). Cephalic Phase of Lipolysis is Impaired in Pancreatic Insufficiency: Role of Gastric Lipase. *Scandinavian Journal of Gastroenterology, 35(2):* 204-211. doi: 10.1080/003655200750024407.

Yu, J. E., Han, S. Y., Wolfson, B., & Zhou, Q. (2018). The Role of Endothelial Lipase in Lipid Metabolism, Inflammation, and Cancer. *Histology and Histopathology, 33(11): 1-10.* doi: 10.14670/HH-11-905.

Yu, W., Lei, Q., Yang, L., Qin, Guohui., Liu, Shasha., Wang, D., Ping, Y., & Zhang, Y. (2021). Contradictory Roles of Lipid Metabolism in Immune Response Within the Tumour Microenvironment. *Journal of Hematology & Oncology, 14(1):* 187. doi: 10.1186/s13045-021-01200-4.

Zambon, A., Deeb, S. S., Pauletto, P., Crepaldi, G., & Brunzell, J. D. (2003). Hepatic Lipase: A Marker for Cardiovascular Disease Risk and Response to Therapy. *Current Opinion in Lipidology, 14(2):* 179-189. doi: 10.1097/00041433-200304000-00010.

Zechner, R., Zimmermann, R., Eichmann, T. O., Kohlwein, S. D., Haemmerle, G., Lass, A., & Madeo, F. (2012). FAT SIGNALS - Lipases and Lipolysis in Lipid Metabolism and Signalling. *Cell Metabolism, 15(3):* 279-291. doi: 10.1016/j.cmet.2011.12.018.

Zhang, H. (2019). Lysosomal Acid Lipase and Lipid Metabolism: New Mechanisms, New Questions, and New Therapies. *Semantic Scholar,* 29: 218-223. https://www.semanticscholar.org/paper/Lysosomal-acid-lipase-and-lipid-metabolism%3A-new-new-Zhang/457ad1f59ae0017c395edd92d1ced732db221a6f.

Chapter 3

Probiotic Lipases: Versatile Catalysts for Sustainable Industries and Therapeutic Innovations

Saraswathy Nagendran[*] and Neha Mudaliar

SVKM's Mithibai College of Arts, Chauhan Institutes of Science and Amrutben Jivanlal College of Commerce and Economics, Mumbai, India

Abstract

Lipases prove to have a wide range of applications, and it is a sustainable option. Their activity of breaking down the triacylglycerol makes it versatile and unique for its action in different industries. It can be extracted from varied sources like plants, animals, and microorganisms, etc. By and large the production of bacterial lipases can be achieved easily through various available technologies. Lipases have different types in accordance with its structure, action, and location, such as Lipoprotein lipase (LPL), Pancreatic phospholipases, Carboxyl ester lipase, Endothelial Lipase (EL). The inhibition of pancreatic lipases using probiotic food products is found to enhance the anti-obesity properties and provoke weight loss due to the fermentation ability of the microorganisms. The probiotic isolates such as *Lactobacillus plantarum* S13, *Lactobacillus delbrueckii* S2, and *Lactobacillus casei* S5 etc., has shown applications in prevention of oxidative rancidity and improve the shelf life of products. It acts as an industrial biocatalyst and also reduces water activity by reversing the reaction. It is involved in diverse bioconversion reactions like interesterification, esterification,

---

[*] Corresponding Author's Email: saraswathynagendran@gmail.com.

In: Lipases and their Role in Health and Disease
Editors: Vasudeo Zambare and Mohd. Fadhil Md. Din
ISBN: 979-8-89113-628-1
© 2024 Nova Science Publishers, Inc.

alcoholysis, acidolysis aminolysis, and hydrolysis. The maturation factor of lipases plays a vital role in their folding and assembly. Genetic modifications in Structured Lipids provide unique characteristics and health benefits that makes further studies essential.

Keywords: lipase, health benefits, probiotic isolates, microbial lipases, nutraceuticals

Introduction

"Probiotics," a term derived from the Greek language, meaning "for life," are considered to be non-pathogenic live microbes that play an essential role in providing numerous health advantages to the host when found to be administrated in sufficient or desired amounts. The term was first coined in 1965 by, 'Lilly and Stillwell,' which referred to the secretion of products by one organism that promotes the growth of another. While Russian immunologist Elie Metchnikoff, a Nobel Laureate in Medicine, was the first to describe the beneficial role of the consumption of probiotics more than a century ago. They are one of the beneficial bacterial species that can be used instead of antibiotics, irradiation, or immunosuppressive therapy for various treatments with minimal effects on gut flora and hence proves to be an attractive option to maintain the microbial equilibrium and prevention of disease (Gupta and Garg, 2009) (Kiousi et al., 2019) (Legesse Bedada et al., 2020) (Reid et al., 2003).

Elie Metchniknoff in 1907 described the concept of traditional probiotics. Based on observations, he suggested that fermented dairy products that include lactic acid (LAB), including yogurt, should be consumed regularly to improve public health and longevity in elderly people. Henceforth, the term is widely linked with beneficial bacteria. Due to wide research and newer technologies, there is an increase in the knowledge regarding microbiota of the human gut and the significance of microbiota imbalances(dysbiosis) in various diseases and syndromes. Probiotics can be considered novel health-promoting strategies which highlight the use of commensal bacteria for restoring the normal balance of flora within the different human ecosystems such as intestinal, vaginal, skin, etc. Extensive studies have been carried out to explore them commercially in different products around the world. Many definitions focus on the viability of microorganisms and their being alive at the site of action, whereas in some instances, the components produced have also shown positive response. The use of non-viable bacteria and their compounds as growth-promoting substrates have opened up new options related to the field (Ouwehand et al., 2002) (Martín and Langella, 2019).

Table 1. List of Bacterial and Fungal genera used in probiotic preparations (Gupta and Garg, 2009)

Sr. No.	Groups	Examples
1	Bacterial Genera	*Bifidobacterium, Lactobacillus, Escherichia, Enterococcus, Bacillus, Streptococcus*
2	Fungal Genera	*Saccharomyces* family

Researchers from various filed have drawn attention to probiotics as well as fermented foods that include fermented milk products. The process of fermentation depends on the types of organisms or mixed microbes such as bacteria, yeasts, and molds that carry out natural fermentation. (Legesse Bedada et al., 2020) Fermented food that consists of bacteria and provides health benefits are called functional foods. The term has an origin in Japan where the concept of food developed for medical benefits has raised during1980s. Furthermore, this food has achieved commercial interest in exploiting the health attributes related to probiotics and significantly contributing to its faster growth and expansion in the market (Stanton et al., 2001).

Health Benefits

The potential benefits of probiotic foods in public health have adequate evidence along with the specific strains that are safe for human use which are stated by The Food and Agriculture Organization of the United Nations and the World Health Organization (WHO) (Reid et al., 2003). The gastrointestinal tract in humans is a complex reservoir that consists of a dynamic and varied population of microorganisms which are considered as the gut microbiota. It contains bacteria that pose a significant influence on the host not during homeostasis but also in diseases. This indicates the presence of prokaryotes ten times higher than that of eukaryotes. The development of gut microbiota usually occurs in the first three years of human life. The host energy balances indirectly, depending on the type of microbiota present which corresponds to the types and proportion of bacteria present in the gut. The intestinal composition of the microbiota is susceptible to change depending on the factors such as stress, age, lifestyle, diet, obesity, intake of medicines, gastrointestinal diseases, etc. the imbalance can lead to intestinal homeostasis disorders that can result in proinflammatory immune responses and can be the route for various diseases including cancer (Śliżewska et al., 2020).

Several bacterial and fungal species till today are used for human health benefits. Among them, both bacterial and fungal species such as Lactobacillus, Bifidobacterium, and *Saccharomyces boulardii* are found to have a positive role as probiotic agents and when consumed in sufficient amounts provide the host the ability to withstand physicochemical conditions in the gastrointestinal tract. Other beneficiary effects of probiotics are demonstrated in some disorders such as allergies, inflammatory bowel disease (IBD), diarrhea, lactose malabsorption, and necrotizing enterocolitis in preterm infants. The mechanism of probiotics majorly is their attachment to the epithelial barrier, which increases adhesion to intestinal mucosa, leading to concomitant inhibition of pathogenic adhesion. It also eliminates pathogens competitively along with the production of antimicrobial substances and causes modulation in dendritic cells. This affects the T cell polarity and causes modulation in the immune system and inflammation (Yousefi et al., 2019).

Probiotics have a wide range of effectiveness ranging from necrotizing enterocolitis, female urogenital infection, infantile diarrhea, *Helicobacter pylori* infections, antibiotic-associated diarrhea, and inflammatory bowel disease to cancer, relapsing *Clostridium difficile* colitis and surgical infections. Beneficial effects on intestinal immunity have been proven by using *Lactobacillus rhamnosus* strain GG, which demonstrates an increase in the number of IgA and other immunoglobulin-secreting cells present in the intestinal mucosa. It also boosts the local release of interferons and increases the antigen uptake in Peyer's patches by facilitating antigen transport to underlying lymphoid cells (Gupta and Garg, 2009). Probiotics have major characteristics such as, 'bile tolerance, acid resistance, mucosal or epithelial cell adhesion, antimicrobial resistance, bile salt hydrolase potential, immunostimulation, antagonistic activity against pathogens, and antimutagenic and anticarcinogenic activities. It is essential to find the potential probiotic strains that can be used with an effective dose of administration (Legesse Bedada et al., 2020).

Food incorporating probiotics, especially fermented foods, contain a wide range of health benefits to humans. They consist of distinctive functional properties such as the production of enzymes, antioxidants, peptides, antimicrobials, and various other properties. These products play a vital role in cancer prevention and treatment. Standard drug usage poses different problems which can be resolved using probiotics which are affordable and safe with a long history of usage. Probiotic bacteria and yeasts are studied and proven to have the ability to abolish carcinogenic toxicity and induction of cancer cell death. Probiotic lactic acid bacteria are one of the important microbes which consist of anticarcinogenic activities. Their effect depends on the strain specificity, metabolic properties, products secreted, and molecules presented. Lactic acid strains have shown effectiveness against cancer both in vitro and in vivo. Probiotic yeasts demonstrate activity against carcinogens and prevent them, using mechanisms such as system alterations in host immunity or through the initiation of apoptotic pathways. It also acts as an antitumorigenic by producing materials that can scavenge mutagens through the gut. They are also used in biotherapeutics as traditionally consuming sufficient amounts of fermented foods with specific doses can be helpful in CRC prevention and treatment (Legesse Bedada et al., 2020).

The gastrointestinal tract acts as a barrier against antigens from microorganisms and food which has led to the establishment of novel therapeutics using probiotics. The indigenous microbiota in the gut depends on the generation of immunohistologic regulation. Probiotics help in creating a non-immunologic gut defense barrier which creates the normalization of increased intestinal permeability and altered gut microecology. Another mechanism of probiotic therapy can be by improvement through immunoglobulin of the intestine's immunologic barrier. An increase in and alleviated inflammatory responses produces gut stabilization. Balance control of proinflammatory and anti-inflammatory cytokines acts as the immune regulation which is the major effect of probiotics indicating that they can be used as tools to increase intestinal inflammation, down-regulate the hypersensitivity reaction, and normalize the gut mucosal dysfunction. The function of probiotics as an immunomodulator can be detected in healthy subjects and recent studies suggest the bacteria needs further characterization for the development of clinical applications for its extended target populations (Isolauri et al., 2001).

Probiotics have also shown beneficial effects on rhinitis, atopic dermatitis, and asthma. The allergic response is one of the increasing global health concerns, specifically for children and people in urban environments. It impairs the quality of life and hence an alternate treatment can be done using probiotics as it has effectiveness in the reduction of symptoms (Lopez-Santamarina et al., 2021).

In recent years the study of probiotics has had a significant increase in research compared to past years, indicating its importance as an emerging field. The consumption of probiotics has also increased rapidly worldwide and developed industries. Further research regarding the standardization of the probiotic strains that can be used to fulfill all the criteria is required for its applications. Production of probiotic foods under good manufacturing practices proven clinically and yet not very reliable products, makes it essential for the development of standards and guidelines which will act as the first step in the development of probiotic products that are effective and legitimate (Reid et al., 2003) (Lu et al., 2021).

The newer approaches, mechanisms, and applications that are under study have the potential to modify scientific understanding along with nutritional and healthcare applications. The microbiome-targeted interventions and other expansions in related fields portend an era of significant change with an anticipated and emerging trend in probiotic sciences with a vision to influence development (Cunningham et al., 2021).

Enzymes as Probiotics

The increase in demand for foods or ingredients that promote health benefits while preventing the onset of diseases is evident. The ingredients that can modify intestinal microbiota and help in modulating organisms can be used for food production and manufacturing process. These ingredients can range from different metabolic products, enzymes, composition, etc. (De Figueiredo et al., 2020). Lactic acid bacteria (LAB) that is used widely as probiotics also produce enzymes such as lipases, amylases, peptidases, ureases, polysaccharide degrading enzymes, phenoloxidases, and esterases. Amylases have a major role in the breakdown or catalyzing of starch and converting them into shorter oligosaccharides by cleavage of α-D 1,4 glycosidic bonds. It can be extracted from various sources such as plants, animals, fungi, and bacteria and used for commercial applications. The hydrolyzed form of starch is much more stable, overcomes the acidic nature, and is used to maintain high temperatures thereby used in fructose syrup production. Some groups of lactic acid bacteria that produce amylase enzyme play a crucial role in the gastrointestinal tract of animals and humans including infants. It has also gained application in commercial industries such as paper, textiles, brewing, and detergents due to its starch liquefaction properties. Not only in industries but its applications are also extended (Figure 1). Few examples of probiotic lactic acid bacteria producing amylolytic activity as *Lactobacillus amylovorus, Lactobacillus plantarum, Lactobacillus manihotivorans*, and *Lactobacillus fermentum*. As these organisms are considered to be non-pathogenic, they can be safely used for production and the lactate obtained as the end product during fermentation can be utilized as a flavoring agent in the food industry.

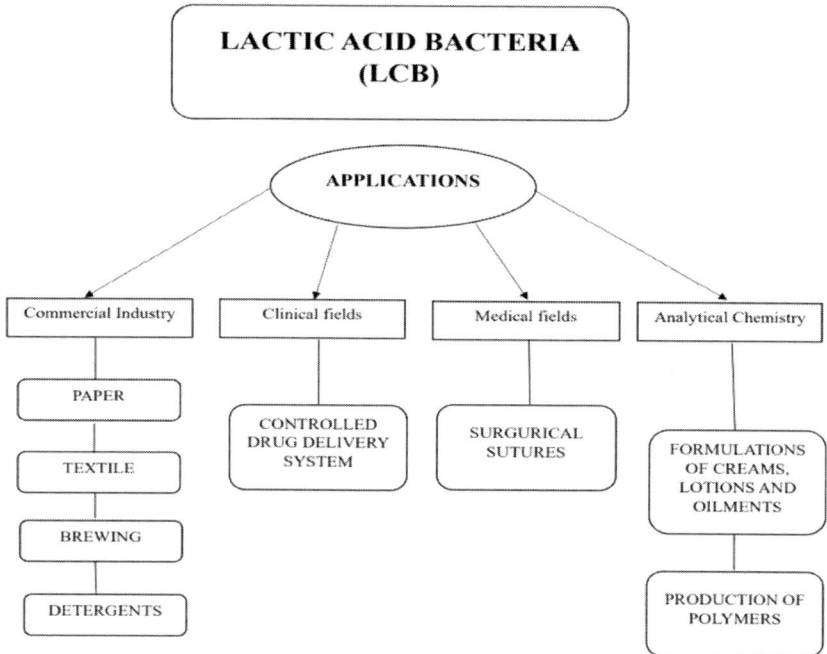

Figure 1. Applications of Lactic Acid Bacteria (LAB).

Recent studies reveal that the *Lactobacillus fermentum* 04BBA19 isolated from soil demonstrated high thermostable α-amylase enzyme. The production of enzymes can be enhanced by using microbial interactions such as *Bacillus amyloliquefaciens* 04BBA15, *Lactobacillus fermentum* 04BBA19, and *S. cerevisiae*. The production of amylase from probiotic lactic acid bacteria is considered generally regarded as safe (GRAS) and the production of lactic acid in the fermentation process makes it industrially economical and has its applications in bioprocessing and fermented foods (Padmavathi et al., 2018).

Other bacteria such as *Bacillus* sp. produce exoenzymes that have been widely used as putative probiotics. When these enzymes were administered in the host i.e., aquaculture it indicated enhanced immunity but the nutritional effects of probiotic bacteria and on digestive enzymes have not yet been evaluated hence the aims for further studies (Ziaei-Nejad et al., 2006).

Food supplements with probiotics or enzymes have been considered as contributing factors for low-nutrient diets and found to improve feed conversion ratio and retention of nutrients. Supplementation of exoenzymes such as protease, amylase, and xylanase along with a maize-soybean meal diet has demonstrated the performance in maintaining low nutrient levels. It also improves the digestibility in diets with lower levels of apparent metabolizable energy (AME). The progress of a probiotic-supplemented nutrient diet is studied and demonstrated in animals and found to be beneficial (Zhi-gang et al., 2014). Other advances in research on extracellular polymeric substances produced by *Lactobacillus plantarum* C70 and *L. plantarum* RJF4 of different origin demonstrates the potential antidiabetic activity by inhibition of α-glucosidase and α-amylase enzymes under in vitro conditions (Angelin and Kavitha, 2020).

Lipases and Its Action

The progress in the field of biotechnology industry opens a completely new advent of enzymology which is used worldwide. Major applications of enzymes in industry depend on various hydrolytic enzymes such as lipases, amylases, proteases, amidases, and esterases. Recently, lipases emerged as one of the key enzymes growing in the field of biotechnology due to their array of applications.

Lipase, being an important industrial biocatalyst, acts as a hydrolytic enzyme, which carries out the cleavage of carboxyl ester bonds that are present in triacylglycerols and liberates fatty acids and glycerol in aqueous conditions. They have natural substrates such as long-chain triacylglycerols which have very low solubility in water and catalyze the reaction at the lipid-water interface. They also possess the unique ability to process the reverse reaction with minimum water activity that leads to alcoholysis, esterification, and acidolysis under micro-aqueous conditions. The property of esterolytic activity besides being lipolytic makes it very diverse with the range of substrates and highly specific as chemo-, regio- and enantioselective catalysts and boosts its use in ester synthesis. Ester synthesis by lipases has applications in numerous fields such as the synthesis of enantiopure pharmaceuticals and nutraceuticals, biodiesel production, fat and lipid modification, resolution of the racemic drugs, and flavor synthesis. Novel technologies like solvent engineering by molecular approaches such as protein engineering and direct evolution and molecular imprinting can be used to improve the catalytic potential of lipases. Other modifications of lipases like surfactant coating to suit the non-aqueous synthesis are also reported. Since the process is unaffected by undesirable side effects it plays a vital role in food processing industries. The stability of the enzyme and its turnover during application conditions are properties of lipases that need to be improved. They work with high specificity for the reactions they catalyze along with being robust and versatile with the wide range of substrates that they can act upon (Gupta et al., 2004) (Rajendran et al., 2009) (Chandra et al., 2020).

Lipases act on the interface between hydrophilic and hydrophobic regions that brings about the characteristic change and distinguishes them from esterases. There are several structures of lipases known to date from different sources and their molecular weight usually ranges from 20,000-60,000. They share the nucleophile-histidine-acidic residue similar to that of serine protease and the catalytic triad is usually a Ser-His-Asp triad or a Ser-His-Glu triad that makes it highly conserved. They consist of disulfide bridges that give it the characteristic feature stability that is essential for its catalytic activity. The hydrophobic residues are surrounded while the catalytic site is usually buried in the molecule. It prevents the enzymes from proteolytic activity by a helical polypeptide structure that acts as a cover and makes the site inaccessible to substrates and solvents. The side of the structure that faces the catalytic site mainly consists of aliphatic hydrophobic side chains whereas the opposite face is usually hydrophilic which is stabilized by protein surface interactions. X-ray studies reveal the phenomenon of interfacial activation is usually associated with the rearrangement of the helical lid structure that increases the hydrophobicity of the surface in the vicinity of the active site and exposes the site. The oil/water interface interaction initiates the opening of the lid which has created a dogmatic view of the mode of action of lipases and is considered a common assumption that the action of lipases requires an interface for its activity. Evidence suggests lipases can undergo inhibition whereas in some cases they are found to be more tolerant to inhibition at the interface. The binding of the hydrophilic surface of the enzyme to the aliphatic side chains of the substrate is very rare in the presence of a hydration layer, for hydrophobic substrates. The folding of the aliphatic side chain of the substrate onto the enzyme's surface is facilitated by an interphase present at the mouth of the active cleft, which in turn, may provide an incomplete hydration layer around the lipase molecule. The local electrostatic patches on the surface of the enzyme may act as further stabilization which provides dipolar attraction to the weakly polarized C-H bonds of the aliphatic side chains (Yahya et al., 1998) (Reyes-Reyes et al., 2022). The knowledge of lipases and their activity that affects the interfacial composition of catalysis requires further study (Reis et al., 2009).

The most common source of lipases is microorganisms followed by plants and mammalian cells. The count on the availability of lipases has significantly increased since the 1980s, which is mainly due to the achievements in the field of cloning and expression of enzymes in different microorganisms. Also, the increase in demand for biocatalysts with different properties such as stability, specificity, pH, and temperature are the reason for the development and increase in the enzyme's potential number. Lipases are classified further into three groups depending on the specificity of the reaction. First is the breakdown of acylglycerol molecules at random sites i.e., non-specific lipases which produce free fatty acids and glycerol along with monoacylglycerols and diacylglycerols as intermediates. The formation of the product is similar to the chemical catalyst, but the difference is the lower product thermo-degradation due to lower temperature reaction in biocatalysis. However, other lipases which is specific (1,3-specific lipases) acts at the position 1 and 3 of the glycerol backbone which causes the particular release of fatty acids from acylglycerol molecules. It is a host of transesterification, esterification (hydrolysis), and interesterification reactions that catalyze enantioselective derivatization of chiral compounds. The hydrolytic activity of lipases is directly correlated to its synthetic activities which were independent of the interesterification activities. Lipases also differ from source to source in their ability to catalyze the same reaction but the performance varies depending on the reaction conditions. During the esterification reaction, the formation of water poses a major issue as it leads to the reverse hydrolysis reaction i.e., ester hydrolysis reaction, and will lead to the formation of hydrolytic side reaction in case of transesterification reaction system. Recently, much more popularity is gained by the choice of lipases from different systems and its study with either solvent-free or hydrophobic organic solvent systems to monitor its performance due to its simplicity and safety in applications (Macrae and Hammond, 1985; Yahya et al., 1998).

The gene family of lipases originally included LPL, Pancreatic lipase, Hepatic lipase (HL). They have evolved from the common ancestor which is indicated by their significant similarity in the amino acid sequences of their proteins. Other two lipolytic enzymes namely endothelial lipase and phosphatidylserine phospholipase A1 have been recently added to this gene family. Synthesis and secretion of HL are majorly carried out by the liver and are bound to proteoglycans and heparan sulfate on the surface of sinusoidal endothelial cells and the external surfaces of microvilli of parenchymal cells in the space of Disse. Functions of HL in lipoprotein metabolism such as chylomicron remnants removal that occurs in parenchymal cells where the presence of an endothelial localization of HL can make the process quite complex therefore demonstrating the relevance to its localization in the liver. HL significantly processes the catalytic activity which contributes to the remodeling of HDL (high-density lipoprotein), VLDL (very-low-density lipoprotein) remnants, and LDL (low-density lipoprotein). Other than its lipolytic activity, HL participates as a ligand and promotes hepatic uptake of lipoproteins with the surface proteoglycans and LDL receptor like-protein, which includes remnants, LDL, and HDL particles. HL plays a vital role in promoting the scavenger receptor B1-mediated uptake of HDL-cholesteryl ester by utilizing both its catalytic and ligand activities. The process of atherosclerosis is possibly influenced by these effects, and the process of reverse cholesterol transport is contributed by HL (Zambon et al., 2003).

The worldwide increase in research for the use of lipases in different biotechnological processes has led to the production of microbial lipases due to the ease in production i.e., by fermentation, availability, and economical. Potential screening of organisms and modification can be beneficial for production and its applications in human welfare (Macrae and Hammond, 1985).

Lipase Production

Lipases hold a wide range of biotechnological applications and act as prominent biocatalysts. The source of enzymes is expanded from plants, animals, and different types of microorganisms. The newer approaches and spectrum of available technology have led to the development in the field of production of lipases and lipase-mediated reactions that directly boost the impact on different industries (Treichel et al., 2010). The improved understanding and study of production biochemistry, the process involved in fermentation, and product recovery has shown a significant increase in the production of enzymes affordably. Enzymes can undergo various catalytic activities and can be multiplied for their action and modified with the help of recent cloning tools that have greatly contributed to expanding the demand for their production (Sharma et al., 2001).

Microbial lipases production has been commercially oriented and found to be more stable in comparison with plants and animals. The microbial lipases provide an opportunity for easy production using fermentation methods. Optimization of organisms to increase the yield and meet the demand for lipase production with various catalytic actions and new sources can be successfully achieved. Microorganisms that produce lipases are present in different sources such as agro-waste, oil-contaminated soil, vegetable oil processing factories, dairy plants, etc. The production of lipases can be enhanced by using natural oils such as vegetable oil, petroleum oil, and coconut oil. Industrially, lipases are important because of their greater potential and large-scale production. For industrial production, research proves that submerged fermentation is a highly effective and promising tool in biotechnology for potentially important enzymes such as lipases, cellulases, and pectinases. Being an appropriate process for developing countries with numerous advantages over solid-state fermentation, such as the resemblance to natural habitat for some microorganisms, reduced energy, better oxygen, fewer operational problems, cost-effectiveness, less effect in downstream processing, compactness of fermentation vessel, higher productivity, lower capital, and recurring expenditure. Many other methods of fermentation are also included in production such as batch, fed-batch, repeated- batch and continuous processes (Treichel et al., 2010).

Different sources of microorganisms are used for the production of lipases such as isolation of strains from petrol spilled oil. The isolation of the enzymes can be initially processed using the pilot plant for determination of activity and further can be optimized for higher temperature, pH, and other factors also with the purification and molecular mass determination (Bharathi et al., 2019). Also, lipase-producing organisms have been found in oil-contaminated soil, industrial waste, oil processing factories, and other habitats. A few of the strains are reported to form various sources, *Bacillus species* from olive oil wastewater, *Geobacillus stearothermophilus* from desert soil samples that can as both lipolytic and thermophilic, and *Stentrophomonas* genus from soil and water habitats. For utilization of these organisms to produce lipases, optimization of every component involved in the process is essential using the conventional method i.e., one factor at a time such that there is one variable while other factors remain constant to enhance the large-scale production. This conventional method of optimization also has disadvantages like time consumption, being expensive, and does not consider the total interactions of different components in the medium which can be overcome by statistical approaches (Hasan-Beikdashti et al., 2012). Not only bacteria but other organisms such as fungi, actinomycetes, and yeasts are also used for the production of lipases. Commercially vital lipase-producing fungi are available as shown in Figure 2. The production of lipase from filamentous fungi varies for the strain, media composition, conditions used for cultivation, temperature, pH, and up to a certain extent the carbon and nitrogen sources (Treichel et al., 2010).

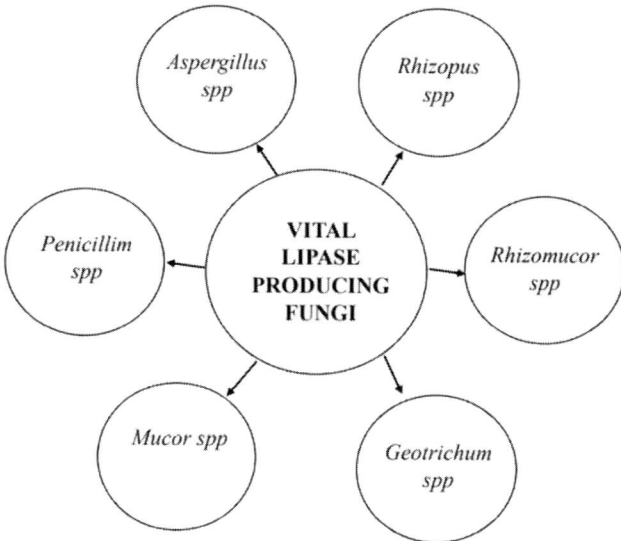

Figure 2. Vital lipase producing fungi (Treichel et al., 2010).

Microbial products can be obtained from free or immobilized cells. The process of immobilization of cells is considered the physical containment or localization of microbial cells. During the industrial process, the continuous use of immobilized cells as catalysts proves to be advantageous in comparison with the batch fermentation method. Immobilization can be carried out in two ways that are either immobilization of the whole cell or immobilization of only the enzyme. However, immobilization of the whole cell has achieved numerous advantages over enzyme immobilization. Research demonstrates the use of various entrapment techniques calcium alginate, polyacrylamide gel, etc. with the effect of other factors such as incubation time, the concentration of sodium alginate, and its reusability of immobilized cells for lipase production (Borrelli and Trono, 2015). Further analysis of lipases and their co-relation with biosurfactants are involved in recent studies that can help in the applications (Colla et al., 2010). The progress in the field of production using different wastes makes the process a more sustainable approach. Oil spills and contamination of water and soil can be used as different substrates for the production of enzymes that not only bring down the cost of production but also play an important role in the remediation of various waste samples and hence majorly the production of lipases is termed as sustainable due to its cost-effective nature that makes it more viable.

Enriched Enzyme Production

To enhance the rate of production, lipid modifications have proved to be a powerful tool to address the above challenges. Changing the fatty acid profile such as the unsaturation level, chain length and positional distribution of natural fats and oils contribute to improving the function and nutrition of structured lipids. Many research papers suggest the modification especially using SLs defined as TAGs along with the modified fatty acid composition and position distribution. Various SLs are discussed based on their commercial interests (Guo et al., 2020).

Lipase modification methods such as preparing them to act in the organic solvent by coating a new surfactant-enzyme complex which leads to the development of lipases with a coat of surfactant. The lipases coated with surfactants are insoluble in water and soluble in most organic solvents due to the solubilization of lipases by hydrophobic tails in the organic media. This modification has a wide application in esterification studies. The reaction catalyzed by the surface-coated lipases has demonstrated an increase in yield compared to that of the direct dispersion or microemulsion system. The solubility of the crude enzyme was extremely low in the reaction, enhanced from 1-9 to 150 folds by activation of lipase absorbed on the surface. Other modification methods, such as modulating the enantioselectivity of organisms, use molecular imprinting techniques. For example, in *Pseudomonas cepacia*, the production was enhanced two-fold using enantioselectivity by coating the surfactant using molecular imprinting methods. The exploitation of synthesis and purification of enantio pure components can boost its applications in food and pharmaceuticals. Modulations in the hydrophobicity of the enzyme used as a biocatalyst can also act as major modifications that can be used for increasing the yield (Rajendran et al., 2009).

Recent research provides vast knowledge regarding molecular modification in the field of enzyme production. Since time, there is an increase in genetic engineering approaches the industrial production of the enzyme has achieved greater success. Modern applications of enzymes cause an elevated demand, which can be met using genetic modifications. Modification in the substrate binding site makes the lipase enzyme specific in terms of its applications. The advent of genetic modification is possible with the detailed study of the existing structure and features of the enzyme. The recombination technique used in microbes can act as a tailoring factor for the production of enzymes with different specificity. The required specificity can be achieved by the process using mutagenesis for appropriate reactions. Development in sequencing provides information regarding the protein sequence which can further be engineered and designed to generate enzymes with key features such as thermostability, oxidative stability, or protease stability by its use.

The transformation of the host used for the expression needs to be analyzed for its expression and before administration has to be considered safe to follow as per the US Food and Drug Administration i.e., termed to be GRAS. Genetic studies reveal that the gene LIP 1 is majorly used for systematic modification for site-directed mutagenesis to achieve functional expression. An open reading frame LIM A was identified downstream of *Pseudomonas cepacia* from gene LIP A. The expression of LIP A occurs only with the presence of LIM A which has been studied in various organisms such as *E. coli, B. subtilis,* and *Streptomyces lividans*. The study encourages the use of plasmid derived from *Geotrichum* species strains containing LIP 1 and LIP 2 gene was used for transformation in yeast strains and high levels of lipase 1 were obtained but the lipase 2 secretion was comparatively higher. The expression of the bacterial lipase gene from *Bacillus subtilis* in yeast has different can be used for different specificities (Rajendran et al., 2009). Both the structural and genetic modifications depend on the requirements of the enzyme and its specificity, the major focus being the production to meet the increasing demand for its applications in various industries.

Applications and Public Health Benefits

Lipases are considered valuable biotechnology enzymes as they pose various applications. Microbial lipases are versatile in nature and have different properties that are used in industries because of the ease of their production. They are widely used as a biocatalyst to manufacture products in the food industry and have applications in the production of fine chemicals. The commercial use of lipase comprises a wide spectrum of applications. Food processing industries majorly demand economic and green technologies to modify fats and oils and use lipases to meet their requirements. Lipases modify the properties of lipids by altering the fatty acid chains and are relatively less expensive compared to other methods. Immobilized lipases have also been sued in the removal of phospholipids in vegetable oils. Due to its ability to catalyze esterification, transesterification, and interesterification reactions in non-aqueous medium thereby making them a versatile choice for their application in paper, pharmaceutical, leather, textile, cosmetic, detergent, and food industries (Ray, 2012) (Rajendran et al., 2009) (Chandra et al., 2020).

Lipases are extensively used in the dairy industry for milk fat hydrolysis. They are used in the alterations of fatty acid chain that enhances the flavor and in producing different varieties of cheese. Cheese ripening, and lipolysis of butter, cream, and fats are also enhanced using lipase enzymes. They are open to a wide range of uses in fruit and vegetable juices, and fat removal processes in meat and fish products. Interesting findings also suggest its addition in noodles which results in a significant change in the characteristics of noodles by making the texture soft even with the presence of low levels of substrate acylglycerols in the formulations. The efficient catalytic reaction of lipases enzyme has also made it significant in pharmaceutical industries. They catalyze synthetic reactions that have been used in the production of life-saving drugs. Creation of optically active homochiral intermediates to synthesize non-steroidal anti-inflammatory drugs such as naproxen, ibuprofen, suprofen, and ketoprox. It is also used in the synthesis of anti-viral, anti-cancer, antibiotics, anti-allergic, alkaloids, vitamins, and anti-arteriosclerotic compounds (Ray, 2012).

Lipases have found their major applications in nutraceuticals where the food components have tremendous health benefits beyond their nutritional value (Reyes-Reyes et al., 2022). Obesityis the main cause of metabolic syndrome and a group of interrelated metabolic conditions, which have increased the risk of cardiovascular diseases. In addition to obesity, health conditions like hypertension, resistance to insulin, non-alcoholic steatohepatitis, and dyslipidemia also require attention. The characteristic feature of dyslipidemia derived from obesity consists of a slight increase in lipoproteins with low density particularly small dense LDL whereas the HDL linked to reduction of cholesterol esters have been found to undergo reduction. The major cause of obesity is the imbalance between energy consumption and its utilization, perticularly due to overconsumption of fats and refined sugars. To absorb the dietary fats, they need to be broken down or hydrolyzed into smaller molecules, a role fulfilled by lipases. Probiotic foods with organisms producing lipases carry out the hydrolyses of fats and prove to give beneficial effects on human health. Pancreatic lipases act as important enzymes, and their activity needs the formation of fat micelles in the intestinal lumen that exerts the activity at the oil-water interface. Fats start to form an emulsion in the mouth that transits to the stomach and is stabilized on reaching the intestine due to the presence of bile salts and other molecules secreted by gastrointestinal fluids. The facts mentioned above demonstrate the urgent need for the development of methods to treat obesity where pancreatic lipases play a key role and process 50-70% of dietary triglycerides converting them to monoacylglycerol and fatty acids which draws a great deal of attention to its application in treatment and improvement in health (Gil-Rodríguez and Beresford, 2021; Nayebhashemi et al., 2023).

Probiotics with lipases have further applications attributed to therapeutics. Evidence suggests further clarity is required for progress in the applications regarding probiotic lipases. Claims suggest that they have benefits in lactose intolerant patients and for weight gain in infants and young animals. The disaccharide lactose leads to severe intestinal distress causing bloating, abdominal pain, and flatulence in individuals with low levels of lactase enzyme. This condition restricts the use of dietary products that can be more severe with age, and leads to limiting consumption of calcium-rich foods that affects the bone density in the elderly. During probiotic formation, the bacteria undergo the fermentation process that leads to the formation of lactase that catalyzes and breaks down lactate to glucose and galactose making their consumption beneficial. It was observed that feeding lactose intolerant subjects fermented milk significantly lowered the hydrogen production in the breath as compared to unfermented milk. The metabolism of lactose before entering the large intestine is indicated by the hydrogen level. A lower level indicates the prior metabolism. They also have beneficial properties in intestinal infection, intestinal immunity, colon cancer, and food allergies (Goldin, 1998).

Conclusion

The tremendous progress in probiotics makes them an essential topic in current research. Amid the demands for healthy foods, probiotics have found their commercial applications along with their potential health benefits. Production of lipases has proved to be easy and environmentally sustainable. The hydrolysis activity of ester bonds present in synthetic plastic, parabens, and insecticides is an emerging aspect to sustain hazardous waste that occurs globally. Another aspect of this, is the generation of high-value-added components by utilizing less energy by enzyme catalysis. Lipases are prime candidates due to their design in therapeutic and diagnostic aids and hence prominently enhancing biotechnological-based production. The progress in genetic modifications, leading to desirable lipases following their specificity is emerging as a boon for industrial applications. Methods of immobilization and protein engineering have broadened the field of production. The consumption of probiotics is still debatable in many aspects due to the microorganism being present and the possibility of pathogenesis in certain conditions. Major research evidence suggests the use as safe and hence focuses on the positive applications and public health benefits if administered in a desired quantity. The effectiveness of lipases as biocatalysts and their vital role in paper, textile, food, dairy, and various other industries has enhanced their applications. It also plays a role in environmental management due to its use in agrochemicals, surfactants, the oleochemical industry, etc. Gut microbes and their interactions with the probiotics significantly help in the construction of microflora that can provide protection from infection and provide immunity to the intestine. They also improve bone health with the increase of beneficial gut flora. The administration of probiotics improves the lactose uptake in lactose intolerant subjects, adding to its benefits. The large number of studies conducted, patents, and research work suggests the importance of probiotic lipases and the growth would be sustained for many years.

References

Angelin, J., & Kavitha, M. (2020). Exopolysaccharides from Probiotic Bacteria and Their Hhealth Potential. *Intetrnational Journal of Biological Macromolecules, 162:* 853–865. doi: 10.1016/j.ijbiomac.2020.06.190.

Bharathi, D., Rajalakshmi, G., & Komathi, S. (2019). Optimization and Production of Lipase Enzyme from Bacterial Srains Isolated from Petrol Spilled Soil. *Journal of King Saud University Science, 31:* 898–901. doi: 10.1016/j.jksus.2017.12.018.

Borrelli, G., & Trono, D. (2015). Recombinant Lipases and Phospholipases and Their Use as Biocatalysts for Industrial Applications. *International Journal of Molecular Science, 16:* 20774–20840. doi: 10.3390/ijms160920774.

Chandra, P., Enespa, Singh, R., & Arora, P. K. (2020). Microbial Lipases and Their Industrial Applications: A Comprehensive Review. *Microbial Cell Factories, 19:* 169. doi: 10.1186/s12934-020-01428-8.

Colla, L. M., Rizzardi, J., Pinto, M. H., Reinehr, C. O., Bertolin, T. E., & Costa, J. A. V. (2010). Simultaneous Production of Lipases and Biosurfactants by Submerged and Solid-State Bioprocesses. *Bioresource Technology, 101:* 8308–8314. doi: 10.1016/j.biortech.2010.05.086.

Cunningham, M., Azcarate-Peril, M. A., Barnard, A., Benoit, V., Grimaldi, R., Guyonnet, D., Holscher, H. D., Hunter, K., Manurung, S., Obis, D., Petrova, M. I., Steinert, R. E., Swanson, K. S., Van Sinderen, D., Vulevic, J., & Gibson, G. R. (2021). Shaping the Future of Probiotics and Prebiotics. *Trends in Microbiology, 29:* 667–685. doi: 10.1016/j.tim.2021.01.003.

De Figueiredo, F. C., De Barros Ranke, F. F. & De Oliva-Neto, P. (2020). Evaluation of Xylooligosaccharides and Fructooligosaccharides on Digestive Enzymes Hydrolysis and as a Nutrient for Different Probiotics and *Salmonella typhimurium*. *LWT, 118:* 108761. doi: 10.1016/j.lwt.2019.108761.

Gil-Rodríguez, A. M., & Beresford, T. (2021). Bile Salt Hydrolase and Lipase Inhibitory Activity in Reconstituted Skim Milk Fermented with Lactic Acid Bacteria. *Journal of Functional Foods, 77:* 104342. doi: 10.1016/J.JFF.2020.104342.

Goldin, B. R. (1998). Health Benefits of Probiotics. *British Journal of Nutrition, 80:* S203–S207. doi: 10.1017/S0007114500006036.

Guo, Y., Cai, Z., Xie, Y., Ma, A., Zhang, H., Rao, P., & Wang, Q. (2020). Synthesis, Physicochemical Properties, and Health Aspects of Structured Lipids: A Review. *Comprehensive Reviews in Food Science and Food Safety, 19:* 759–800. doi: 10.1111/1541-4337.12537.

Gupta, R., Gupta, N., & Rathi, P. (2004). Bacterial Lipases: An Overview of Production, Purification and Biochemical Properties. *Applied Microbiology and Biotechnolgy, 64:* 763–781. doi: 10.1007/s00253-004-1568-8.

Gupta, V., & Garg, R. (2009). Probiotics. *Indian Journal of Medical Microbiology, 27:* 202–209. doi: 10.4103/0255-0857.53201.

Hasan-Beikdashti, M., Forootanfar, H., Safiarian, M. S., Ameri, A., Ghahremani, M. H., Khoshayand, M. R., & Faramarzi, M. A. (2012). Optimization of Culture Conditions for Production of Lipase by a Newly Isolated Bacterium *Stenotrophomonas maltophilia*. *Journal of Taiwan Institute of Chemical Engineering, 43:* 670–677. doi: 10.1016/J.JTICE.2012.03.005.

Isolauri, E., Sütas, Y., Kankaanpää, P., Arvilommi, H., & Salminen, S. (2001). Probiotics: Effects on Immunity. *American Journal of Clinical Nutrition, 73:* 444s–450s. doi: 10.1093/ajcn/73.2.444s.

Kiousi, D., Karapetsas, A., Karolidou, K., Panayiotidis, M., Pappa, A., & Galanis, A. (2019). Probiotics in Extraintestinal Diseases: Current Trends and New Directions. *Nutrients, 11,* 788. doi: 10.3390/nu11040788.

Legesse Bedada, T., Feto, T. K., Awoke, K. S., Garedew, A. D., Yifat, F. T., & Birri, D. J. (2020). Probiotics for Cancer Alternative Prevention and Treatment. *Biomedicine and Pharmacotherapy, 129:* 110409. doi: 10.1016/j.biopha.2020.110409.

Lopez-Santamarina, A., Gonzalez, E. G., Lamas, A., Mondragon, A. D. C., Regal, P., & Miranda, J. M. (2021). Probiotics as a Possible Strategy for the Prevention and Treatment of Allergies. A Narrative Review. *Foods, 10:* 701. doi: 10.3390/foods10040701.

Lu, K., Dong, S., Wu, X., Jin, R., & Chen, H. (2021). Probiotics in Cancer. *Frontiers in Oncology, 11:* 638148. doi: 10.3389/fonc.2021.638148.

Macrae, A. R., & Hammond, R. C. (1985). Present and Future Applications of Lipases. *Biotechnology and Genetic Engineering Reviews, 3:* 193–218. doi: 10.1080/02648725.1985.10647813.

Martín, R., & Langella, P. (2019). Emerging Health Concepts in the Probiotics Field: Streamlining the Definitions. *Frontiers in Microbiology, 10:* 1047. doi: 10.3389/fmicb.2019.01047.

Nayebhashemi, M., Enayati, S., Zahmatkesh, M., Madanchi, H., Saberi, S., Mostafavi, E., Mirbzadeh Ardakani, E., Azizi, M., & Khalaj, V. (2023). Surface Display of Pancreatic Lipase Inhibitor Peptides by Engineered *Saccharomyces boulardii*: Potential as an Anti-Obesity Probiotic. *Journal of Functional Foods, 102:* 105458. doi: 10.1016/j.jff.2023.105458.

Ouwehand, A. C., Salminen, S., & Isolauri, E. (2002). Probiotics: An Overview of Beneficial Effects. In: *Lactic Acid Bacteria: Genetics, Metabolism and Applications,* Siezen, R. J., Kok, J., Abee, T., & Schasfsma, G. (Eds.), Springer Netherlands, Dordrecht, 279–289. doi: 10.1007/978-94-017-2029-8_18.

Padmavathi, T., Bhargavi, R., Priyanka, P. R., Niranjan, N. R. & Pavitra, P. V. (2018). Screening of Potential Probiotic Lactic Acid Bacteria and Production of Amylase and Its Partial Purification. *Journal of Genetic Engineering and Biotechnolgy, 16:* 357–362. doi: 10.1016/j.jgeb.2018.03.005.

Rajendran, A., Palanisamy, A., & Thangavelu, V. (2009). Lipase Catalyzed Ester Synthesis for Food Processing Industries. *Brazilian Archives of Biology and Technology, 52:* 207–219. doi: 10.1590/S1516-89132009000100026.

Ray, A. (2012). Application of Lipase in Industry. *Asian Journal Pharmacy and Technology, 2(2):*33-37. https://ajptonline.com/AbstractView.aspx?PID=2012-2-2-1.

Reid, G., Jass, J., Sebulsky, M. T., & McCormick, J. K. (2003). Potential Uses of Probiotics in Clinical Practice. *Clinical Microbiological Reviews, 16:* 658–672. doi: 10.1128/CMR.16.4.658-672.2003.

Reis, P., Holmberg, K., Watzke, H., Leser, M. E. & Miller, R. (2009). Lipases at Interfaces: A Review. *Advances in Colloid and Interface Science, 147–148:* 237–250. doi: 10.1016/j.cis.2008.06.001.

Reyes-Reyes, A. L., Valero Barranco, F., & Sandoval, G. (2022). Recent Advances in Lipases and Their Applications in the Food and Nutraceutical Industry. *Catalysts, 12:* 960. doi: 10.3390/catal12090962.

Sharma, R., Chisti, Y., & Banerjee, U.C. (2001). Production, Purification, Characterization, and Applications of Lipases. *Biotechnology Advances, 19:* 627–662. doi: 10.1016/s0734-9750(01)00086-6.

Śliżewska, K., Markowiak-Kopeć, P., & Śliżewska, W. (2020). The Role of Probiotics in Cancer Prevention. *Cancers, 13:* 20. doi: 10.3390/cancers13010020.

Stanton, C., Gardiner, G., Meehan, H., Collins, K., Fitzgerald, G., Lynch, P. B., & Ross, R. P. (2001). Market Potential for Probiotics. *American Journal of Clinical Nutrition, 73:* 476s–483s. doi: 0.1093/ajcn/73.2.476s.

Treichel, H., De Oliveira, D., Mazutti, M. A., Di Luccio, M., & Oliveira, J. V. (2010). A Review on Microbial Lipases Production. *Food and Bioprocess Technology, 3:* 182–196. doi: 10.1007/s11947-009-0202-2.

Yahya, A. R. M., Anderson, W. A., & Moo-Young, M. (1998). Ester Synthesis in Lipase-Catalyzed Reactions. *Enzyme and Microbial Technology, 23:* 438–450. doi: 10.1016/S0141-0229(98)00065-9.

Yousefi, B., Eslami, M., Ghasemian, A., Kokhaei, P., Salek Farrokhi, A., & Darabi, N. (2019). Probiotics Importance and Their Immunomodulatory Properties. *Journal of Cellular Physiology, 234:* 8008–8018. doi: 10.1002/jcp.27559.

Zambon, A., Bertocco, S., Vitturi, N., Polentarutti, V., Vianello, D., & Crepaldi, G. (2003). Relevance of Hepatic Lipase to the Metabolism of Triacylglycerol-Rich Lipoproteins. *Biochemial Society Transactions, 31:* 1070–1074. doi: 10.1042/bst0311070.

Zhi-gang, T., Naeem, M., Chao, W., Tian, W., & Yan-min, Z. (2014). Effect of Dietary Probiotics Supplementation with Different Nutrient Density on Growth Performance, Nutrient Retention and Digestive Enzyme Activities in Broilers. *Journal of Animal & Plant Sciences, 24(5):*1309-1315.

Ziaei-Nejad, S., Rezaei, M. H., Takami, G. A., Lovett, D. L., Mirvaghefi, A. R., & Shakouri, M. (2006). The Effect of *Bacillus* spp. Bacteria Used as Probiotics on Digestive Enzyme Activity, Survival, and Growth in the Indian White Shrimp *Fenneropenaeus indicus*. *Aquaculture, 252:* 516–524. doi: 10.1016/j.aquaculture.2005.07.021.

Chapter 4

Bio-Surfactants: Green Synthesis Using Lipases and Role in Human Health

Benu Arora[*]
Department of Applied Chemistry and Environmental Studies,
Bhagwan Parshuram Institute of Technology,
Guru Gobind Singh Indraprastha University, New Delhi, India

Abstract

Bio-surfactants are a group of environmentally friendly surfactants that are in great demand worldwide because of the numerous advantages that they offer over their chemical counterparts. They categorise into various groups of structurally diverse compounds, the most commonly studied and applied members being the glycolipids. Deriving glycolipids from renewable resources rather than from petroleum-based ones is an eco-friendly method. There are two main ways of obtaining glycolipids: one involving microbial fermentation of renewable agro-industrial wastes and the other being the enzymatic route that uses sugars and fatty acids (or their esters) as the substrates and isolated enzymes (mostly immobilized) as the catalyst. Herein, we focus on the enzymatic route for the production of bio-surfactants. The initial part of this chapter gives an account of the application of bio- catalysis in the synthesis of bio-surfactants and few effective techniques that are used in the improvement of this class of biotransformation. Some of successful applications of glycolipids (or Sugar Fatty Acid Esters, SFAEs) include the ones in food, cosmetic, petroleum, pharmaceutical and textile industries. Also significant is their use in the agricultural and waste water, soil treatment sectors. However, it is the biomedical applications, which are of greater

---

[*] Corresponding Author's Email: benu.arora01@gmail.com.

In: Lipases and their Role in Health and Disease
Editors: Vasudeo Zambare and Mohd. Fadhil Md. Din
ISBN: 979-8-89113-628-1
© 2024 Nova Science Publishers, Inc.

importance in the context of human health. The antimicrobial and antibiofilm forming activities possessed by this class of compounds are useful in the prevention of several microbial infections. Additionally, some of the glycolipids have been successfully employed as drug delivery and anti-cancer agents, thereby increasing their therapeutic importance. The second half of this chapter describes some of these important applications in the healthcare industry.

Keywords: lipases, bio-surfactants, glycolipids, sugar, organic solvents, fatty, biocatalyst, anti-microbial activities, therapeutic agents

Introduction

Surfactants are chemical compounds that make their presence felt in almost every walk of life from food, cosmetics, pharmaceuticals and agrochemicals to petrochemicals, metallurgy, environmental remediation and so on (De et al., 2015; Kralova and Sjöblom, 2009). The global annual production of surfactants has seen a sharp rise over the last few decades. However, most of the surfactants are synthesized using petroleum, which is a limited resource. Hence, the focal point of research in recent years has been the search for alternative resources that can be used to produce surfactants. Bio-surfactants are promising alternatives to the petrochemical surfactants as the former are sustainably derived from renewable resources (Singh and Saini, 2013; Rebello et al., 2013; Marcelino et al., 2020). Additionally, the chemical surfactants are non-biodegradable, leading to various health as well as environmental problems (Nagtode et al., 2023). The increased legal and societal pressures for consumer products to be biodegradable coupled with consumer awareness about the sustainability factor and safety of the products are some of the major reasons that have driven the research behind the production of these next-generation eco-friendly surfactants. In fact, the last two decades have seen a sharp rise in the number of research papers published on bio-surfactants (Singh and Saini, 2013). Between 2020 and 2027, the global green surfactant market is expected to expand at a Compound Annual Growth Rate (CAGR) of 5.7%, mainly due to the increased demand for products sourced from renewable resources (White et al., 2013).

Bio-surfactants are surface active agents produced as secondary metabolites of microorganisms. They possess a number of desirable properties like foaming, emulsifying, dispersing, solubilizing, wetting as well as self-assembling abilities. Some of the important advantages that they offer include their environmentally benign nature, ability to undergo biodegradation, and performance under extreme temperature and pH conditions (Santos et al., 2016; Grüninger et al., 2019). The important classes of bio-surfactants are glycolipids, lipopeptides, lipopolysaccharides and phospholipids. Glycolipids, also commonly referred to as SFAEs form the best-known group within bio-surfactants (Desai and Banat,1997). The immense structural diversity of glycolipids again helps their categorization into a number of subclasses. Of these, the more frequently studied ones (in terms of applications) are the rhamnolipids, trehalolipids, sophorolipids and mannosylerythritol-lipids.

Glycolipids are synthesized *de novo* via microbial fermentation of renewable feedstock and then extracted, separated via physical and chemical processes. Such production methods have their own set of advantages as well as limitations (Ljunger et al., 1994; Grüninger et al., 2019). A detailed discussion about these fermentation methods is not within the scope of this chapter. However, it is worth mentioning over here that the major limitation of this method is the high cost of the overall process. This is attributed mainly to the low yield of bio-surfactants in the culture media and the high cost of downstream processing (Singh and Saini, 2013). An alternate way to synthesize SFAEs within the realm of the chemical laboratory, aided by bio-catalysts involves the combining of a suitable sugar molecule (mono/disaccharide) and a fatty acid (free or esterified) via esterification/transesterification reaction (Therisod and Klibanov, 1987; Neta et al., 2015). This route not only overcomes the limitations of the microbial fermentation method but also offers certain distinct advantages such as high reaction yields, high regioselectivity and the possibility to reuse the enzyme without major loss of activity (Hollenbach et al., 2020; Saab-Rincon et al., 2023). Thus, enzymatic synthesis of SFAEs has emerged as a promising alternative in many cases. However, there are few limitations of this approach that researchers are trying to overcome. The search for greener reaction solvents, designing better performing and/or reusable biocatalyst designs and extending the enzymatic route synthesize the esters of oligosaccharides as well are few significant areas of current research (Saab-Rincon et al., 2023).

Due to their non-toxicity and high emulsifying capacity, SFAEs are extensively used in the food processing industry (Campos et al., 2013; Neta et al., 2015). These chemicals are also highly demanded in the cosmetics industry due to their skin-friendliness and compatibility (Khan and Rathod, 2015; Gayathiri et al., 2022; Sarubbo et al., 2022). Owing to their antimicrobial, anti-adhesive and enzyme-inhibiting properties, bio-surfactants find a variety of applications in the fields of medicine and pharmacy (Markande et al., 2021). Bio-surfactants are also the preferred alternatives to the synthetic pesticides, fertilizers and other crop protecting agents. Hence, there are numerous reports on their being successfully used in the field of agriculture (Sachdev and Cameotra, 2013; Farias et al., 2014; Singh et al., 2018). Bio-surfactants also hold a lot of promise as green alternatives for the synthesis of nanoparticles (Mulligan, 2005; Markande et al., 2021). Another very significant application of bio-surfactants that has far reaching effects on human health is soil and waste water remediation (Banat et al., 2010; Sarubbo et al., 2015). While there are a large number of reviews that very well summarize these applications and many others, the focus of this chapter is on the few important medical applications of bio-surfactants. Figure 1 summarizes the range of topics covered in the present chapter.

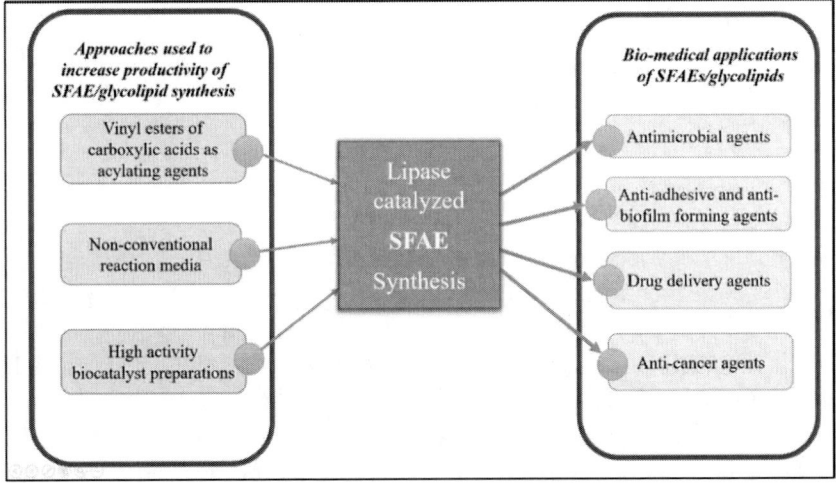

Figure 1. Scope of this chapter.

Glycolipids: Structure-Activity Correlation

Glycolipids or SFAEs are amphiphilic molecules, possessing a hydrophilic head group comprised of mono-, di-, or oligosaccharides and a hydrophobic tail consisting of one or more alkyl chains containing different number of methylene units. This dual character enables them to partition themselves at the interface between different phases such as solid- liquid or liquid-gas or between liquids of different polarities such as oil-water. Consequently, they reduce the surface tension of a liquid and interfacial tension between two liquids. An important aspect related to bio-surfactants is their lower Critical Micelle Concentration (CMC) values as compared to their chemical counterparts (Sobrinho et al., 2014). CMC is the concentration of the surfactant above which organized aggregates of the surfactant molecules, known as micelles are formed, and corresponds to the point of attainment of lowest stable surface tension by the tension-active agent. Possession of low CMC values by bio-surfactants enables their utilization and exploitation in different important areas such as cosmetics, food processing, pharmaceuticals, bioremediation of oil spills, removal of toxic metals etc. (Gayathiri et al., 2022).

Some of the structural features of the SFAE molecule that are worth mentioning are the length of the alkyl chain, extent of acylation of sugar and type of linkage between the sugar moiety. In aqueous solution, the CMC value of a SFAE is directly influenced by the length of the fatty acid chain (Ducret, 1995; Polat et al., 1997). Since the hydrophilic head composed of sugar is relatively small in size, it appears to have influence in this regard (Ferrer et al., 2002). The degree of acylation of the sugar moiety is also found to affect the surface activity of glycolipids. An increased number of hydrophobic tails diminishes the SFAE's surface activity due to decrease in the efficiency of aggregation at the surface. This in turn has been attributed to the opening up of the surfactant structure (Husband et al., 1998). Another variable that impacts the surfactant efficiency is the spacing between the head and the tail which is determined by the type of linkage between the two. The linking bond can either be an amine, amide, ether or ester functionality (Stubenrauch, 2001). The type of linkage is also found to have a strong influence on the stability and thereby on the biodegradability of the bio-surfactant. Increased interfacial activity is seen as the alkyl chain moves farther from the head (Razafindralambo et al., 2009, Razafindralambo et al., 2012).

Lipases: Biocatalysts for the Synthesis of SFAEs (Glycolipids)

The synthesis of SFAEs requires coupling of a sugar and a fatty acid (free or esterified) in the presence of a suitable catalyst. Sugars being polyhydroxy compounds, present the challenge of regioselectivity modifying one (or more) of the several hydroxyl groups that are part of their structure. The conventional chemical methods of SFAE synthesis require harmful catalysts and solvents, elevated temperatures and also produce multiple reaction products. Such reactions barely conform to the basic principles of green chemistry. Moreover, some of the by-products may be allergic and possibly carcinogenic (Arcos et al., 1998; Gumel et al., 2011). Enzyme catalyzed synthesis of glycolipids provides a green alternative that is more acceptable to the environmentally conscious consumers (Divakar and Manohar, 2007). In fact, biocatalysis is considered one of the most environmentally friendly routes for the synthesis of important chemical compounds. Enzyme catalyzed transformations require ambient temperature and pressure conditions. Moreover, extra purification steps are minimally required as most biocatalysts are stereo and/or regio-selective (Sheldon and Woodley, 2018).

HLB (Hydrophilic-Lipophilic Balance) value of a surfactant is measure of its hydrophilicity (or lipophilicity) (Griffin, 1949). In case of SFAEs, the HLB value is determined by the type of acyl group, extent of substitution and degree of polymerization of the carbohydrate (Zheng et al., 2015). Enzyme catalyzed synthetic route also makes it possible to synthesize a wide spectrum of bio-surfactants by changing the head and/or tail part of their molecular structure (Pedersen et al., 2002). While nature offers us a wide range of hydrolases such as lipases, proteases, glucosidases, and esterases from a variety of plants, animals, and microbes, it is lipases that have been most frequently used for the purpose of producing SFAEs (Hayes, 2011).

Lipases (triacyl glycerol hydrolases, EC 3.1.1.3) are type of hydrolases. Their normal function in aqueous medium is to bring about the hydrolysis of carboxylic ester bonds of fats (Kapoor and Gupta, 2012). It is now well established that lipases are also capable of bringing about the synthesis of ester bonds in non-aqueous media, a phenomenon which is widely referred to as condition promiscuity (Hult and Bergland, 2007; Arora et al., 2014). Although, synthesis of the ester bond does not need water as such, yet for optimum enzyme activity it is essential to maintain enzyme hydration and therefore a certain minimum amount of water is a must even in non-aqueous reaction systems (Zaks and Klibanov, 1998). Over the years, synthetic organic chemists have exploited lipases by making them catalyse a variety of reactions such as esterification, transesterification, interesterification, alcoholysis etc. (Majumder et al., 2006; Arora, 2021). This is due to a reversal in the lipase activity from hydrolytic to condensation. Most of the lipases used for glycolipid synthesis are sourced from the microbial world and are used in the immobilized form (Zago et al., 2021).

Table 1. Some illustrative examples of SFAE synthesis

Sr. No	Sugar source	Acyl donor	Enzyme formulation	Solvent system	Additional reaction condition(s)	Reference
1	Glucose	Vinyl decanoate	Novozymes 435 (immobilized CALB)	Deep Eutectic Solvents (DES)	Beed mill apparatus	Hollenbach et al., 2022
2	Fructose	Myristic acid	Lipozyme RMIM, Novozymes 435	Oleic acid	Low solvent media	Cabezas et al., 2022
3	Sucrose	Fatty acid vinyl esters (FAVEs)	Lipozyme TLIM	t-butanol /pyridine (1: 1, v/v)	-	Zhu et al., 2022
4	D-Glucose	Lauric acid	Novozymes 435	2-methyl-2-butanol (2M2B); 2-methyl-tetrahydrofuran and 2-methyl-tetrahydrofuran-3-one	Green solvents used as reaction medium	Vuillemin et al., 2022
5	Glucose, Xylose	Lauric acid, methyl- and vinyl-laurate	Lipozyme 435	2M2B	Membrane assisted solvent recovery	Martinez-Garcia et al., 2021
6	Glucose, Sucrose, Fructose	Vinyl laurate	Novozymes 435	ILs	Supersaturated sugar solution used	Shin et al., 2019
7	Trehalose	Lauric acid, ethyl laurate	Novozymes 435	-	Solvent free conditions	Ogawa et al., 2019
8	Maltohep-taose	Fatty acids such as lauric acid, myristic acid	*Candida antarctica* Lipase A	Dimethyl Sulphoxide (DMSO)/ t-butanol (10/90 v/v)	-	Nguyen et al., 2019
9	Sugars obtained from honey and agave syrup	FAVEs	Immobilized *Candida antarctica* Lipase B	Honey and agave syrup	-	Siebenhaller et al., 2018
10	Lactose	Oleic acid	Lipozyme® (immobilized from *Mucor miehei*)	Water-saturated toluene (50% v/w)	-	Perinelli et al., 2018
11	Sucrose	Vinyl caprate	*Candida rugosa* lipase	DMSO	Non-aqueous Biphasic system	Inprakhon, 2017

Table 1. (Continued)

Sr. No	Sugar source	Acyl donor	Enzyme formulation	Solvent system	Additional reaction condition(s)	Reference
12	Glucose	Fatty acids such as hexanoic acid, palmitic acid	Immobilized *Candida antarctica* lipase	DMSO/2M2B (8:2)	-	Ren and Lamsal, 2017
13	Glucose	Stearic acid	MNPs prepared from Fe_3O_4 and *Rhizobium oryzae*)	Isooctane	-	Sebatini et al., 2016
14	Glucose	Vinyl laurate	Novozymes 435	Ionic Liquid (IL)/ mixture of IL, Volatile Organic Solvent	-	Lin et al., 2015
15	Fructose	Lauric acid	Novozymes 435	IL/2M2B cosolvent system	-	Li et al., 2015
16	Galactose	Amino acids, fatty acid acyl chlorides	Lipozyme TL IM	*t*-butanol	-	An et al., 2015
17	Glucose, maltose	Vinyl propionate/ ethyl acrylate	Novozymes 435	Dioxane	Two-step process combining lipase and CGTase used to produce oligo-saccharide esters	Ayres et al., 2014
18	Maltodextrin	Decanoic acid, lauric acid, palmitic acid	*Thermomyces lanuginosus* lipase	DMSO	-	Udomrati and Gohtani, 2014
19	Oligofructose	Fatty acids such as stearic acid, caprylic acid	Novozymes 435	DMSO/ *t*-butanol (10/90 v/v)	-	Van Kempen et al., 2014
20	Maltose	Linoleic acid	Novozymes 435 and lyophilized powders of few other lipases	Either no solvent or a mixture of two ionic liquids	-	Fischer et al., 2013

The application of lipases for catalyzing SFAE synthesis has been a topic of extensive study for the past many years (Markande et al., 2021; Sarubbo et al., 2022). However, the last few years have seen the development of newer strategies and technologies that have overcome many of the limitations that were faced by the earlier researchers. Several research papers and reviews discuss the various dimensions of lipase catalyzed SFAE synthesis (Nieto et al., 2021; Barros et al., 2023). Only some of these advancements are being covered in this chapter. Table 1 summarizes some of the latest examples of lipase catalyzed SFAE synthesis from the last ten years.

Choice of Reaction Media

One of the most critical factors that determines choice of the reaction medium while designing SFAE synthesis is optimizing substrate's solubility in the reaction medium without compromising on the lipase stability (Gumel et al., 2011). Since the polarities of sugars and fatty acids vary a lot, so achieving sufficiently high concentrations of both substrates in the same medium becomes challenging. An additional challenge comes because of the conflicting solubility of sugars and activity of lipases in a given organic solvent. Sugars being polyhydroxy compounds are appreciably soluble in highly polar organic solvents (Moye, 1972). An effective way of co-relating the enzyme activity with the solvent polarity as expressed by log P has been proposed by Laane et al., (1987). Here P is the partition coefficient of the given component in the standard octanol-water two-phase system. Polar solvents with log P < 2 are generally considered a poor choice for enzyme catalysis, as they strip off the essential water from the enzyme molecules. On the other hand, the non-polar solvents with log P > 4 are the best choice for obtaining high enzyme activity. However, as mentioned earlier, the problem with SFAE synthesis is that a given organic solvent's suitability is being governed by two opposing factors.

The significance of solvent polarity in the context of lipase catalyzed esterification/ transesterification reactions of sugars can be understood by taking the example of a simple monosaccharide, glucose. Product conversion/ yield and regioselectivity are the two important parameters that need to be optimized while synthesizing glucose esters. On one end of the solvent spectrum are the non-polar or hydrophobic organic solvents like hexane and toluene, which despite supporting lipase activity are considered non-suitable for SFAE synthesis due to the poor solubility of glucose in them. On the other end of the spectrum are highly polar solvents like DMSO and DMF which although show appreciable solubility for glucose yet deactivate the lipase and therefore are not suitable for SFAE synthesis (Degn and Zimmermann, 2001). The product conversion and regioselectivity for the transesterification reaction between D-glucose and vinyl acetate mediated by *Candida antarctica* lipase B in organic solvents of different polarity has been studied in detail (Arora, 2021). As shown in Figure 2 this reaction leads to the formation of two products: the monoacetate (major product) and the diacetate (minor product).

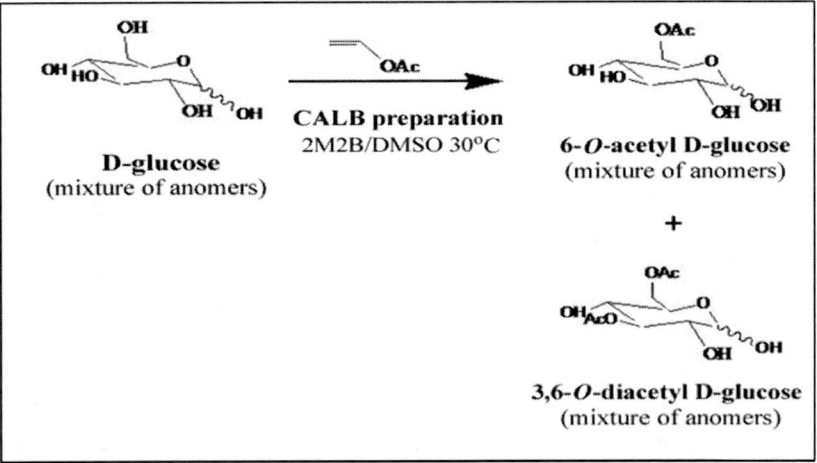

Figure 2. Glucose fatty acid ester synthesis catalysed by *Candida antarctica* Lipase B (CALB).

The significance of solvent polarity in the context of lipase catalyzed esterification/ transesterification reactions of sugars can be understood by taking example of a simple monosaccharide, glucose. Product conversion/ yield and regioselectivity are the two important parameters that need to be considered over here. On one end of the solvent spectrum are the highly hydrophobic solvents such as hexane and toluene, that are considered non-suitable for SFAE synthesis due to the extremely limited solubility of glucose in these solvents. On the other end of the spectrum are highly polar solvents like DMSO and DMF which although show appreciable solubility for glucose yet deactivate the lipase and therefore are not suitable for SFAE synthesis (Degn and Zimmermann, 2001).

In acetone and *t*-butanol (moderately polar solvents): high product conversions have been obtained (owing to reasonably high glucose solubility):

but the reaction exhibited poor regioselectivity. On the other hand, in DMF (highly polar solvent): although the product conversions were poor, yet there was an exclusive formation of the monoacetate. Thus, simultaneously obtaining high product conversions and high regioselectivity in any one organic solvent is a challenge that researchers have been facing. Consequently, over the last two decades, they have developed several strategies and techniques that enable the attainment of these dual targets. Many of these approaches are also aimed at making the overall synthetic process more acceptable from the perspective of Green Chemistry (Khan and Rathod, 2018; Vuillemin et al., 2022). The next few sections of this chapter cover three such approaches: one involving the use of unconventional reaction media, second one involving FAVEs as the acylating agents and the third one describing the use of High Activity Biocatalyst Preparations (HABPs).

Employing Unconventional Reaction Media for SFAE Synthesis

As mentioned earlier, lipases exhibit condition promiscuity in nonaqueous solvents. This aspect of lipases in fact, opens up the opportunity for medium engineering (Zaks and Klibanov, 1988). Medium engineering encompasses reaction methodologies such as use of mixed organic solvent systems, ionic liquids and deep eutectic solvents.

Use of Mixed Organic-Solvent Systems
A simple yet effective way to improve the sugar solubility while maintaining the lipase stability is the application of mixed organic solvents as reaction media. Generally, in the context of SFAE synthesis, solvent mixtures composed of t-butanol or 2M2B along with 5-30% (v/v) of a polar cosolvent such as DMSO, DMF or pyridine are used. An important benefit of employing mixed solvent systems is the possibility to gradually increase the hydrophilicity of the medium by a stepwise increment in the percentage of polar co-solvent. This in turn helps in retaining optimum enzyme activity as there is no drastic change in the enzyme's immediate vicinity. This approach results in extremely good product yields coupled with high regioselectivity during the synthesis of SFAEs (Ferrer et al., 2000).

Ren and Lamsal used solvent mixtures composed of 2M2B and DMSO for esterifying glucose with fatty acids like such as palmitic acid, lauric acid and hexanoic acid (Ren and Lamsal, 2017). Different 6-O-acylglucose esters were prepared in a range of organic solvent mixtures and evaluated as emulsifiers (Liang et al., 2018). SFAEs of disaccharides such as sucrose, maltose, lactose and lactulose have also been synthesized by employing mixed organic solvents as the reaction medium (Ferrer et al., 2005a; Reyes-Duarte et al., 2005; Chávez-Flores et al., 2017).

For biocatalytic processes utilizing organic solvents as reaction medium to be economically feasible, solvent recovery and recycling are very important. In fact, the high cost and environmental issues related to most of the organic solvents are often the reasons for search of better solvent systems. Solvent recycling offers a possible solution to this. A recent work dealing with glucose and xylose esters utilized nanofiltration using organic solvent resistant membranes for this purpose (Martinez-Garcia et al., 2021). This energy efficient technique also makes the overall process environmentally more acceptable. More research on similar lines could enable the improvement of economics as well as sustainability of bio-surfactant production.

Use of Ionic Liquids (ILs)

Ionic liquids (ILs) are organic salts with a melting point less than 100°C. They have emerged as the green solvents of choice for biocatalysis (Sheldon et al., 2002). Apart from being non-flammable and non-volatile, ionic liquids also impart a stabilizing effect on the enzymes (Elgharbawy et al., 2021). Many of the ionic liquids have been successfully employed as the reaction solvent in the production of bio-surfactants mediated by enzymes (Park and Kazlauskas, 2001; Kim et al., 2003; Ganske and Bornscheuer, 2005). Some excellent reviews cover the work done in this area in the last two decades (Galonde et al., 2012; da Cruz Silvério and Rodrigues, 2020). Like in the case of organic solvents, one of the major limitations while employing ILs in such reactions is striking a balance between the enzyme stability and sugar solubility. Most of the ILs that have high sugar solubility are detrimental to the enzyme activity (Forsyth et al., 2002; Liu et al., 2005; van Rantwijk et al., 2006). One of the possible solutions for this problem involves a water mediated method in which an IL that favours enzyme stability is mixed with an aqueous sugar solution followed by vacuum evaporation of water from the mixture. This results in the creation of a supersaturated sugar solution in the ionic liquid and is found to result in better product yields (Lee et al., 2008). Another successful approach to circumvent this problem is one that is similar to the mixed-organic solvent approach. A mixture of two ILs is prepared: one that is hydrophilic and favours high lipase activity while another that is hydrophobic and favours high lipase stability (Ha et al., 2010; Mai et al., 2014).

Use of ILs is limited not just to the lipase catalyzed SFAE synthesis involving mono- and di-saccharides. The synthesis of starch palmitate catalyzed by lipases has been successfully carried out in mixed ILs (Lu et al., 2012). The action of lipases also enables lignocellulosic biomass valorisation in ILs (Šibalić et al., 2023). In another interesting example involving polysaccharides, esterified cellulose nanocrystals (E-CNCs) have been prepared in a binary IL system (Zhao et al., 2017). Although ILs have proved to be the solvents of choice in several places, there are some aspects of ILS such as their high price, safety and environmental benignity that still remain questionable (Vekariya, 2017).

Use of Deep Eutectic Solvents (DESs)

Deep Eutectic Solvents are formed from a eutectic mixture of Lewis or Brønsted acids and bases. These can be composed of different anionic and cationic species. These solvents are recognized as analogues of ILs (Smith et al., 2014). As they consist of supramolecular structures formed by extensive hydrogen bonding between hydrogen bond donors and acceptors, DESs are liquids at room temperature. In the context of biocatalysis, some of the important advantages that they offer include non-toxicity, biodegradability, non-volatility, enzyme stabilizing effects and the ability to dissolve high substrate concentrations (Gutiérrez et al., 2009; Dai et al., 2015; Chen et.al, 2019). However, one of the major constraints encountered while applying these solvents for biocatalysis is their highly viscous nature (Dai et al., 2013; Procentese et al., 2015). A relatively new yet effectual way to overcome this limitation is mechanoenzymology (Pérez-Venegas and Juaristi, 2021). Hollenbach and co-workers carried out effective synthesis of SFAEs by applying the bead mill apparatus (Hollenbach et al., 2022). In another interesting application of DESs for enzymatic synthesis of glycolipids, the Beechwood carbohydrates obtained from lignocellulose were employed as the substrates, demonstrating a sustainable protocol to obtain SFAEs (Siebenhaller et al., 2017).

Utilizing Fatty Acid Vinyl Esters (FAVEs) for Acylating Sugars

Esters of vinyl alcohol and fatty acids, commonly referred to as vinyl esters are known to be excellent acylating agents in lipase catalysed transesterification reactions of sugars. These enol esters have also been used extensively for carrying out the regio- and/or stereo-selective acylations of glycerol and other secondary alcohols (Bornscheuer and Yamane, 1995; Ferreira et al., 2012). Since the vinyl alcohol generated (as a by-product) during the transesterification reaction undergoes quick tautomerization to a volatile product i.e., acetaldehyde, it makes the reaction irreversible. Thus, the biotransformation proceeds at a faster rate, is more selective and offers easier product isolation. While the literature is full of examples of successful syntheses of SFAEs using FAVEs, this chapter lists a few selected ones wherein the SFAEs so synthesized have subsequently been evaluated for some application(s).

It is worth mentioning that sucrose esters prepared using FAVEs are considered safe food additives if the overall exposure to these is within the ADI of 40 mg/kg bw/day (Younes et al., 2023). Zhang and co-workers evaluated the surface-active properties and antimicrobial activities of various disaccharide monoesters acylated with FAVEs (Zhang et al., 2014). Lactulose, a simple prebiotic sugar underwent regioselective transesterifications using vinyl laurate (Chávez-Flores et al., 2017). Sucrose monolaurate synthesized using vinyl laurate proved to be excellent anti-bacterial against four pathogenic bacteria (Shao et al., 2018). A number 6-O-acylglucose esters were prepared regioselectivity by using various FAVEs. The SFAEs so synthesized not only possessed good emulsifying properties but were also non-toxic towards the cell lines tested by the authors, making them suitable for application as food emulsifiers (Liang et al., 2018). Zhu et al., have synthesized sucrose esters using vinyl acetate, vinyl palmitate, vinyl laurate and vinyl stearate and shown that these SFAEs possessed anti-microbial as well as anti-cancer activities (Zhu et al., 2022).

Employing High Activity Biocatalyst Preparations (HABPs)

Most of the bio-transformations carried out in organic solvents use either lyophilized powders or immobilized form(s) of the enzyme which are "straight from the vendor" preparations. However, there is ample evidence suggesting that such enzymes preparations are not suitable most of the times and need to be replaced by the high activity biocatalyst preparations (Roy and Gupta, 2004; Hudson et al., 2005). Details about the specific preparation methods and applications of each of these have been discussed at several places (Kapoor and Gupta, 2012; Arora, 2021).

Figure 3 shows the basic scheme for the preparation of Enzyme Precipitated and Rinsed with Organic Solvent (EPROS): Cross-linked

Enzyme Aggregates (CLEAs): Protein Coated Micro-Crystals (PCMCs) and Cross-linked Protein Coated Micro-Crystals (CLPCMCs) that are the most commonly utilized HABPs. Some of these biocatalyst designs are prepared by "pre-treating" the enzymes with water miscible organic solvents that helps in the "washing away" of excess water associated with the enzyme, accompanied by simultaneous enzyme precipitation (Solanki and Gupta, 2008). Precipitation of the enzyme alone (i.e., w/o a co-solute) using a water miscible organic solvent followed by rinsing with the same solvent results in formation of EPROS (Solanki and Gupta, 2011). CLEA formation involves crosslinking the precipitate in situ by using a suitable cross-linker such as glutaraldehyde (Schoevaart et al., 2004).

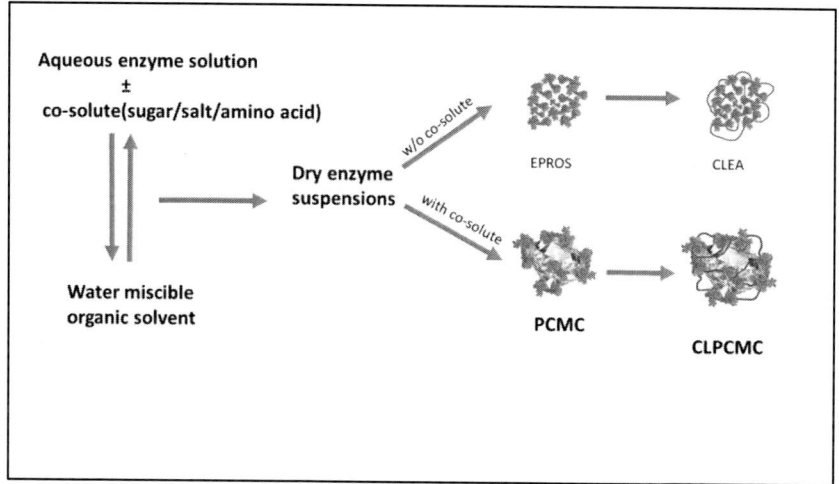

Figure 3. Schematic representation of preparation of HABPs.

Formation of PCMCs involves the co-precipitation of the enzyme along with a specific core material, that is comprised of a low-molecular weight compound such as a sugar, a salt or an amino acid (Kreiner et al., 2001). Cross-linking of PCMCs generates CLPCMCs. In many instances, CLPCMCs gave much better results than PCMCs (Shah et al., 2008; Raita et al., 2011). Some of these designs such as CLEAs also offer the additional advantage of catalyst recycling (Sheldon, 2007).

The utilization of such high activity lipase preparations has also shown promising results in the case of SFAE synthesis. HABPs of CALB such as CLEAs, PCMCs and cross-linked protein coated microcrystals CLPCMCs have been employed for carrying out the fast and regioselective acylation of glucose (Arora, 2021). CLEAs of the lipase from *Thermomyces lanuginosus* have been applied for the synthesis of sucose-6-acetate, an intermediate in the synthesis of sucrose (Yang et al., 2012). In a synergistic approach combining nanotechnology and biocatalysis, magnetic CLEAs of CALB were prepared and evaluated for the synthesis of biosurfactant sucrose- 6'-monopalmitate (López et al., 2014). Although, HABPs of lipases hold a lot of potential for the improvement of SFAE syntheses, this is one area where only limited work has been reported till date.

Bio-Medical and Therapeutic Applications of Bio-Surfactants

As mentioned earlier, applications of bio-surfactants have touched many of the aspects of human life. The impact of these applications on human health either directly or indirectly is especially significant. Of all these, one important application that impacts human health the most is the utilization of biosurfactants in the field of biomedicine and therapeutics. Two important functions are performed by these molecules that make them valuable for medical applications: one is their ability to partition at the interfaces, thereby affecting the adhesion properties of microorganisms while the other is their ability to bring about cell lysis by disrupting the cell membranes (Sánchez et al., 2006).

A large number of research have proved the antimicrobial, anti-adhesive, anticarcinogenic, antiviral and immune-modulating properties of various classes of biosurfactants (Naughton et al., 2019; Liu et al., 2020). The purpose of this chapter is to highlight some of the illustrative examples pertaining to only glycolipids. Among the glycolipids, rhamnolipids and sophorolipids are the more commonly studied bio-surfactants in the field of biomedicine (Thakur et al., 2021; de Oliveira et al., 2015). Rhamnolipids are mainly produced from *Pseudomonas aeruginosa* whereas, sophorolipids are produced by yeasts.

Antimicrobial Activity

As the instances of antibiotic resistance increase globally, there is an emerging trend for bio-surfactants to be used as the safer alternatives. In fact, over the last few years, research has shifted to ways to utilize biosurfactants as adjuvants with current antibiotics (Liu et al., 2020). This stems from the fact that many of them possess bactericidal, bacteriostatic, biofilm inhibition and biofilm disruption abilities. An excellent paper by Rodrigues et.al. describes several significant clinical applications of the antimicrobial activity of bio-surfactants (Rodrigues et al., 2006).

Several mechanisms have been postulated to explain the mode of action of glycolipids against micro-organisms (Meincken et al., 2005; Sana et al., 2018). This chapter covers only a brief account of the antibacterial, antifungal and antiviral activities of glycolipids. Benincasa *et.al.* (2004) found that the rhamnolipids (RL) possessed good antimicrobial behaviour against several pathogenic bacterial and phytopathogenic fungal species. Similarly, the rrhamnolipids produced by *Pseudomonas aeruginosa* AT10 from soybean oil refinery wastes were found to exhibit excellent antifungal properties (Abalos et al., 2001). A low molecular weight glycolipid produced by *Pseudozyma fusiformata* (Order Ustilaginales) has been reported to be an effective fungicide in acidic conditions (Golubev et al., 2001; Kulakovskaya et al., 2003). The novel marine Staphylococcus saprophyticus SBPS 15 produces a glycolipid that not only possesses great temperature and pH stability but also holds a lot of promise as an antimicrobial agent against several bacterial and fungal strains (Mani et al., 2016). Mannosylerythritol lipid (MEL): a yeast glycolipid has also been shown to possess antibacterial activity (Kitamoto et al., 1993).

Sophorolipids are known to be active against HIV (Shah et al., 2005). Similarly, the rhamnolipids produced by a *Pseudomonas* sp. strain are known to be active against herpes simplex virus types 1 and 2 (Remichkova et al., 2008). Nitschke *et.al.* found that the growth of several bacteria and fungi could be curtailed by using rhamnolipids produced from soybean oil waste (Nitschke et al., 2010). Flocculosin is a glycolipid produced by *Pseudomonas flocculosa*. This biosurfactant possesses impressive antibacterial as well as antifungal activities against several pathogenic strains (Mimee et al., 2005; Mimee et al., 2009).

Notably, few examples of antimicrobial activity of SFAEs synthesized chemically. Recently, the esters of monosaccharides such as glucose, xylose and arabinose were prepared enzymatically. These molecules showed activity against several gram-positive bacteria. (Giorgi et al., 2022). The esterification of levoglucosan mediated by lipases in a continuous flow reaction has been carried out and the esters so obtained found to possess good antibacterial activities (do Nascimento et al., 2022). Disaccharide monoesters with medium chain fatty acids have also been shown to have antimicrobial activity against several pathogens (Shao et al., 2018).

Anti-Adhesive and Anti-Biofilm Forming Activity

An inherent property of bio-surfactants is to form aggregates at phase boundaries and interfaces. These surface-active compounds influence the original surface properties by forming an organic- conditioning film. The details about the ability of bio-surfactants to help in microbial adhesion and deadhesion have been described in literature (Neu, 1996). Biofilms are structured, three-dimensional consortia formed by most of the microorganisms when present in contact with solid surfaces in a wet environment. These biofilms are a serious cause of chronic infections and therefore pose major threats in the field of medicine (De Giani et al., 2021). The anti-adhesion and biofilm disruption properties of glycolipids makes them promising candidates for the curtailing the spread of pathogens on surfaces. It is worthwhile to emphasise over here the practical utility of this aspect of bio-surfactants.

One of the simplest and most effective ways to prevent the spread of pathogenic microorganisms is to pre-coat the desired solid surface with a suitable bio-surfactant. E-Silva *et.al.* (2017) showed that rhamnolipids can disrupt the biofilms formed by *Staphylococcus aureus* which is an important pathogen. Maťátková and coworkers have shown that *Pseudomonas aeruginosa* strains produce glycolipids that are capable in combating the biofilm formation by the pathogen *Trichosporon cutaneum* (Maťátková et al., 2021).

In an important application of chemically synthesized SFAEs as antimicrobial and anti-biofilm agents, Campana and co-workers prepared a series of SFAEs and demonstrated their ability to disrupt the biofilms formed by several food- borne pathogens (Campana et al., 2019).

Applications as Drug Delivery Agents

An upcoming application of bio-surfactants in the field of therapeutics is their ability to act as agents to deliver drugs to targeted areas. In-fact, this is a significant field that connects nanotechnology and microbial biotechnology. Many bio-surfactants, particularly rhamnolipids have been exploited for synthesizing nanoparticles, in an eco-friendly manner (Kumar et al., 2010; Saikia et al., 2013).

In another approach related to the above mentioned one, rhamnolipids were used as nano-carriers for delivering drugs to the skin in an ex-vivo system (Müller et al., 2017). The researchers loaded these glycolipid nanoparticles with hydrophobic drugs such as Nile red, dexamethasone and tacrolimus for their studies. Nanoparticles with 30% drug loading were prepared. As detected by fluorescence microscopy, the drug Nile red could be efficiently delivered to isolated human skin and no toxic effects were seen at concentration higher than the CMC values.

Transdermal Therapeutic Systems (TTS): also known as transdermal patches are used to deliver drugs to the body via skin absorption. This form of drug administration has several advantages such as immediate visual confirmation of drug administration once the patch is applied, controlled rate of drug absorption and avoiding the first-pass metabolism by the liver. In an interesting application of SFAEs, sucrose esters were used for a drug delivery in TTS (Csóka et al., 2007). The authors tested several sucrose esters for the release of metoprolol tartrate, the model drug used. Another elegant work involving sucrose esters studied the gel forming properties of these molecules and successfully used them for the *in-vitro* release of drugs such as paracetamol (Szűts et al., 2010).

Anticancer Activity

Owing their diverse biophysical and biochemical properties, glycolipids have emerged as important candidates for cancer chemotherapy and as safe agents for anti-cancer drug delivery. Among glycolipids, maximum research in this area has been done with sophorolipids. Natural as well as modified sophorolipids possess anti-cancer activity against cancers of the pancreas, breast, cervical, colon, liver and brain. Several mechanistic explanations including inhibition of proliferation, induction of apoptosis and membrane disruption have been proposed to explain this action of sophorolipids (Miceli et al., 2022). Li et al., have studied the anticancer potential of sophorolipids for the treatment of human cervical cancer (Li et al., 2017). Recently, acidic sophorolipid preparations were shown to decrease the proliferation of colorectal cancer cells (Callaghan et al., 2022). Rhamnolipids have also proved to be potent anti-cancer agents. Recently, it was demonstrated that the mono-rhamnolipids produced by *Pseudomonas aeruginosa* damaging the colorectal cancer cells (Twigg et al., 2022).

Semkova et al., (2021) studied the anti-cancer activity of mono- and di-rhamnolipids against human breast cancer. Mishra and co-workers have made similar findings. According to their work, rhamnolipids are active against the breast cancer MDA-MB-231 cell line (Mishra et al., 2021). In another important application of rhamnolipids, Yi and co-workers used rhamnolipid nanoparticles loaded with pheophorbide and injected those intravenously into the SCC7 tumour bearing mice. These nanoparticles showed high accumulation in the tumour tissue. When irradiated by laser, complete *in vivo* tumour suppression by photodynamic therapy was observed (Yi et al., 2019).

In a significant example of application of SFAEs, sucrose esters were prepared enzymatically and shown to have antiproliferative effect on six tumour cell lines (Zhu et al., 2022). Similarly, esters of D-maltotriose were found to exert strong cytotoxic effects against various cancer cell lines (Ferrer et al., 2005b).

Conclusion

Bio-surfactants or SFAEs have emerged as promising multi-functional molecules for the future. A part of current research is focussed on to newer ways of synthesizing them enzymatically by using raw materials, preferably sourced from renewable feedstocks. While most of the research is focused on the enzymatic synthesis and applications of SFAEs produced using mono- and di-saccharides, there are only a few reports on the synthesis of oligosaccharide derivatives. Another big issue concerning SFAE synthesis is designing reaction conditions that are 'greener' and therefore more acceptable. The traditional uses of bio-surfactants, particularly glycolipids included bioremediation of certain kinds of pollutants and utilization in the petroleum, agriculture and food industries. It is now well evident that the applications of these molecules in emerging areas such as nanomedicine, anti-cancer therapeutics and pharmaceuticals are on the rise. Though, most of these reports talk about the naturally occurring rhamnolipids and sophorolipids there are few that discuss the biomedical applications of SFAEs synthesized enzymatically using lipases. All this points towards the need for greater research towards searching and /or designing newer strategies such as developing better biocatalyst designs, green reaction solvents, renewably sourced raw materials etc. These will help improve enzymatic production of

SFAEs, thereby fuelling more extensive application of bio-surfactants in areas that improve human health. Obviously, the role of lipases is paramount in this regard.

References

Abalos, A., Pinazo, A., Infante, M. R., Casals, M., Garcia, F., & Manresa, A. (2001). Physicochemical and Antimicrobial Properties of New Rhamnolipids Produced by *Pseudomonas aeruginosa* AT10 from Soybean Oil Refinery Wastes. *Langmuir, 17(5):* 1367-1371. doi:10.1021/la0011735.

An, D., Zhao, X., & Ye, Z. (2015). Enzymatic Synthesis and Characterization of Galactosyl Monoesters. *Carbohydrate Research, 414*: 32-38. doi:10.1016/j.carres.2015.05.011.

Arcos, J. A., Bernabe, M., & Otero, C. (1998). Quantitative Enzymatic Production of 6-O-Acylglucose Esters. *Biotechnology and Bioengineering, 57(5):* 505-509. doi:10.1002/(sici)1097-0290(19980305)57:5%3c505::aid-bit1%3e3.0.co;2-k.

Arora, B. (2021). Highly Efficient Synthesis of Glucose Fatty Acid Esters Catalyzed by High Performance Lipase Preparations. *Asian Journal of Chemistry, 33*: 2489-2497. doi:10.14233/ajchem.2021.23391.

Arora, B., Mukherjee, J., & Gupta, M. N. (2014). Enzyme Promiscuity: Using the Dark Side of Enzyme Specificity in White Biotechnology. *Sustainable Chemical Processes, 2*: 1-9. doi: https://doi.org/10.1186/s40508-014-0025-y.

Ayres, B. T., Valença, G. P., Franco, T. T., & Adlercreutz, P. (2014). Two-Step Process for Preparation of Oligosaccharide Propionates and Acrylates using Lipase and Cyclodextrin Glycosyl Transferase (CGTase). *Sustainable Chemical Processes, 2*: 1-8. doi: . https://doi.org/10.1186/2043-7129-2-6.

Banat, I. M., Franzetti, A., Gandolfi, I., Bestetti, G., Martinotti, M. G., Fracchia, L., Smyth, T. J. & Marchant, R. (2010). Microbial Biosurfactants Production, Applications and Future Potential. *Applied Microbiology and Biotechnology, 87*, 427-444. DOI: https://doi.org/10.1007/s00253-010-2589-0.

Barros, E. L., Rebelatto, E. A., Mayer, D. A., Wancura, J. H., & Oliveira, J. V. (2023). Lipase-catalyzed Production of Sugar Esters in Pressurized Fluid Media: A Review. *Chemical Engineering and Processing-Process Intensification*, 109480. doi:10.1016/j.cep.2023.109480.

Benincasa, M., Abalos, A., Oliveira, I., & Manresa, A. (2004). Chemical Structure, Surface Properties and Biological Activities of the Biosurfactant Produced by *Pseudomonas aeruginosa* LBI From Soapstock. *Antonie Van Leeuwenhoek, 85:* 1-8. doi: 10.1023/B:ANTO.0000020148.45523.41.

Bornscheuer, U. T., & Yamane, T. (1995). Fatty Acid Vinyl Esters as Acylating Agents: A New Method for the Enzymatic Synthesis of Monoacylglycerols. *Journal of the American Oil Chemists' Society, 72(2):* 193-197. doi: 10.1007/BF02638899.

Cabezas, J. T., Waglay, A., & Karboune, S. (2022). Lipase-Catalyzed Synthesis of Fructosyl Myristic Acid Esters as Biosurfactants in a Low Solvent Media:

Optimization of the Bioconversion. *Food Bioscience*, *50*, 102026. doi:10.1016/j.fbio.2022.102026.

Callaghan, B., Twigg, M. S., Baccile, N., Van Bogaert, I. N., Marchant, R., Mitchell, C. A., & Banat, I. M. (2022). Microbial Sophorolipids Inhibit Colorectal Tumour Cell Growth in Vitro and Restore Haematocrit in Apcmin+/−Mice. *Applied Microbiology and Biotechnology, 106(18)*:6003-6016. doi:10.1016/j.fbio.2022.102026.

Campana, R., Merli, A., Verboni, M., Biondo, F., Favi, G., Duranti, A., & Lucarini, S. (2019). Synthesis and Evaluation of Saccharide-Based Aliphatic and Aromatic Esters as Antimicrobial and Antibiofilm Agents. *Pharmaceuticals, 12(4):* 186. doi:10.3390/ph12040186.

Campos, J. M., Montenegro Stamford, T. L., Sarubbo, L. A., de Luna, J. M., Rufino, R. D., & Banat, I. M. (2013). Microbial Biosurfactants as Additives for Food Industries. *Biotechnology progress, 29(5):* 1097-1108. doi:10.1002/btpr.1796.

Chávez-Flores, L. F., Beltran, H. I., Arrieta-Baez, D., & Reyes-Duarte, D. (2017). Regioselective Synthesis of Lactulose Esters by *Candida antarctica* and *Thermomyces lanuginosus* Lipases. *Catalysts, 7(9):* 263. doi:10.3390/catal7090263.

Chen, Z., Jacoby, W. A., & Wan, C. (2019). Ternary Deep Eutectic Solvents for Effective Biomass Deconstruction at High Solids and Low Enzyme Loadings. *Bioresource Technology, 279*: 281-286. doi:10.1016/j.biortech.2019.01.126.

Csóka, G., Marton, S., Zelko, R., Otomo, N., & Antal, I. (2007). Application of Sucrose Fatty Acid Esters in Transdermal Therapeutic Systems. *European Journal of Pharmaceutics and Biopharmaceutics, 65(2):* 233-237. doi:10.1016/j.ejpb.2006.07.009.

da Cruz Silvério, S. I., & Rodrigues, L. R. M. (2020). Biocatalysis in Ionic Liquids: Enzymatic Synthesis of SFAEs. In *Nanotechnology-Based Industrial Applications of Ionic Liquids* (pp. 51-79). Cham: Springer International Publishing. DOI: 10.1007/978-3-030-44995-7_4.

Dai, Y., van Spronsen, J., Witkamp, G. J., Verpoorte, R., & Choi, Y. H. (2013). Natural Deep Eutectic Solvents as New Potential Media for Green Technology. *Analytica Chimica Acta, 766*: 61-68. doi:10.1016/j.aca.2012.12.019.

Dai, Y., Witkamp, G. J., Verpoorte, R., & Choi, Y. H. (2015). Tailoring Properties of Natural Deep Eutectic Solvents with Water to Facilitate Their Applications. *Food Chemistry, 187*: 14-19. doi:10.1016/j.foodchem.2015.03.123.

De Giani, A., Zampolli, J., & Di Gennaro, P. (2021). Recent Trends on Biosurfactants with Antimicrobial Activity Produced by Bacteria Associated with Human Health: Different Perspectives on Their Properties, Challenges, and Potential Applications. *Frontiers in Microbiology, 12*: 655150. doi:10.3389/fmicb.2021.655150.

de Oliveira, M. R., Magri, A., Baldo, C., Camilios-Neto, D., Minucelli, T., & Celligoi, M. A. P. C. (2015). Sophorolipids A promising biosurfactant and it's applications. *International Journal of Advanced Biotechnology and Research, 6(2):* 161-174.

De, S., Malik, S., Ghosh, A., Saha, R., & Saha, B. (2015). A Review on Natural Surfactants. *RSC Advances, 5(81):* 65757-65767. doi:10.1039/c5ra11101c.

Degn, P., & Zimmermann, W. (2001). Optimization of Carbohydrate Fatty Acid Ester Synthesis in Organic Media by a Lipase from *Candida antarctica*. *Biotechnology and bioengineering, 74(6):* 483-491. doi:10.1002/bit.1139.

Desai, J. D., & Banat, I. M. (1997). Microbial Production of Surfactants and Their Commercial Potential. *Microbiology and Molecular biology reviews, 61(1)*: 47-64. doi:10.1128%2Fmmbr.61.1.47-64.1997.

Divakar, S., & Manohar, B. (2007). Use of Lipases in the Industrial Production of Esters. In *Industrial enzymes: Structure, function and applications* (pp. 283-300). Dordrecht: Springer netherlands. doi:10.1007/1-4020-5377-0_17.

do Nascimento, M. A., Vargas, J. P., Rodrigues, J. G., Leão, R. A., de Moura, P. H., Leal, I. C., Bassut, J., de Souza, R. O. M. A, Wojcieszak, R. & Itabaiana, I. (2022). Lipase-Catalyzed Acylation of Levoglucosan in Continuous Flow: Antibacterial and Biosurfactant Studies. *RSC Advances, 12(5)*: 3027-3035. doi:10.1039/d1ra08111j.

Ducret, A., Giroux, A., Trani, M., & Lortie, R. (1995). Enzymatic Preparation of Biosurfactants from Sugars or Sugar Alcohols and Fatty Acids in Organic Media Under Reduced Pressure. *Biotechnology and Bioengineering, 48(3)*: 214-221. doi:10.1002/bit.260480308.

E-Silva, S. S., Carvalho, J. W. P., Aires, C. P., & Nitschke, M. (2017). Disruption of *Staphylococcus aureus* Biofilms using Rhamnolipid Biosurfactants. *Journal of Dairy Science, 100(10)*: 7864-7873. doi:10.3168/jds.2017-13012.

Elgharbawy, A. A., Moniruzzaman, M., & Goto, M. (2021). Facilitating Enzymatic Reactions by Using Ionic Liquids: A Mini Review. *Current Opinion in Green and Sustainable Chemistry, 27*: 100406. doi:10.1016/j.cogsc.2020.100406.

Farias, C. B., Ferreira Silva, A., Diniz Rufino, R., Moura Luna, J., Gomes Souza, J. E., & Sarubbo, L. A. (2014). Synthesis of Silver Nanoparticles Using a Biosurfactant Produced in Low-Cost Medium as Stabilizing Agent. *Electronic Journal of Biotechnology, 17(3)*: 122-125. doi:10.1016/j.ejbt.2014.04.003.

Ferreira, H. V., Rocha, L. C., Severino, R. P., & Porto, A. L. (2012). Syntheses of Enantiopure Aliphatic Secondary Alcohols and Acetates by Bioresolution with Lipase B from *Candida antarctica*. *Molecules, 17(8)*: 8955-8967. doi:10.3390/molecules17088955.

Ferrer, M., Comelles, F., Plou, F. J., Cruces, M. A., Fuentes, G., Parra, J. L., & Ballesteros, A. (2002). Comparative Surface Activities of Di-and Trisaccharide Fatty Acid Esters. *Langmuir, 18(3)*: 667-673. doi:10.1021/la010727g.

Ferrer, M., Cruces, M. A., Plou, F. J., Bernabé, M., & Ballesteros, A. (2000). A Simple Procedure for the Regioselective Synthesis of Fatty Acid Esters of Maltose, Leucrose, Maltotriose and N-Dodecyl Maltosides. *Tetrahedron, 56(24)*: 4053-4061. doi:10.1016/S0040-4020(00)00319-7.

Ferrer, M., Soliveri, J., Plou, F. J., López-Cortés, N., Reyes-Duarte, D., Christensen, M., Copa-Patino, J.L. & Ballesteros, A. (2005a). Synthesis of Sugar Esters in Solvent Mixtures by Lipases from *Thermomyces lanuginosus* and *Candida antarctica* B, and Their Antimicrobial Properties. *Enzyme and Microbial Technology, 36(4)*: 391-398. doi:10.1016/j.enzmictec.2004.02.009.

Ferrer, M., Pérez, G., Plou, F. J., Castell, J. V., & Ballesteros, A. (2005b). Antitumour Activity of Fatty Acid Maltotriose Esters Obtained by Enzymatic Synthesis. *Biotechnology and Applied Biochemistry, 42(1)*: 35–39. doi:10.1042/ba20040122.

Fischer, F., Happe, M., Emery, J., Fornage, A., & Schütz, R. (2013). Enzymatic Synthesis of 6-and 6'-O-Linoleyl-A-D-Maltose: From Solvent-Free to Binary Ionic Liquid

Reaction Media. *Journal of Molecular Catalysis B: Enzymatic*, *90*, 98-106. doi:10.1016/j.molcatb.2013.01.019.

Forsyth, S. A., MacFarlane, D. R., Thomson, R. J., & Von Itzstein, M. (2002). Rapid, Clean, and Mild O-Acetylation of Alcohols and Carbohydrates in an Ionic Liquid. *Chemical Communications*, *(7):* 714-715. doi:10.1039/b200306f.

Galonde, N., Nott, K., Debuigne, A., Deleu, M., Jerôme, C., Paquot, M., & Wathelet, J. P. (2012). Use of Ionic Liquids for Biocatalytic Synthesis of Sugar Derivatives. *Journal of Chemical Technology & Biotechnology*, *87(4):* 451-471. doi:10.1002/jctb.3745.

Ganske, F., & Bornscheuer, U. T. (2005). Lipase-Catalyzed Glucose Fatty Acid Ester Synthesis in Ionic Liquids. *Organic Letters*, *7*(14): 3097-3098. doi:10.1021/ol0511169.

Gayathiri, E., Prakash, P., Karmegam, N., Varjani, S., Awasthi, M. K., & Ravindran, B. (2022). Biosurfactants: Potential and Eco-Friendly Material for Sustainable Agriculture and Environmental Safety—A Review. *Agronomy*, *12(3):* 662. doi:10.3390/agronomy12030662.

Giorgi, V., Botto, E., Fontana, C., Della Mea, L., Vaz Jr, S., Menéndez, P., & Rodríguez, P. (2022). Enzymatic Production of Lauroyl and Stearoyl Monoesters of D-Xylose, L-Arabinose, and D-Glucose as Potential Lignocellulosic-Derived Products, and Their Evaluation as Antimicrobial Agents. *Catalysts*, *12(6):* 610. doi:10.3390/catal12060610.

Golubev, W. I., Kulakovskaya, T. V., & Golubeva, E. W. (2001). The Yeast *Pseudozyma fusiformata* VKM Y-2821 Producing an Antifungal Glycolipid. *Microbiology*, *70:* 553-556. doi: 10.1023/A:1012356021498.

Griffin, W. C. (1949). Classification of surface-active agents by "HLB." *The Journal of the Society of Cosmetic Chemists*, *1*, 311-326.

Grüninger, J., Delavault, A., & Ochsenreither, K. (2019). Enzymatic Glycolipid Surfactant Synthesis from Renewables. *Process Biochemistry*, *87:* 45-54. doi:10.1016/j.procbio.2019.09.023.

Gumel, A. M., Annuar, M. S. M., Heidelberg, T., & Chisti, Y. (2011). Lipase Mediated Synthesis of SFAEs. *Process Biochemistry*, *46(11):* 2079-2090. doi:10.1016/j.procbio.2011.07.021.

Gutiérrez, M. C., Ferrer, M. L., Mateo, C. R., & del Monte, F. (2009). Freeze-Drying of Aqueous Solutions of Deep Eutectic Solvents: A Suitable Approach to Deep Eutectic Suspensions of Self-Assembled Structures. *Langmuir*, *25(10):* 5509-5515. doi:10.1021/la900552b.

Ha, S. H., Hiep, N. M., Lee, S. H., & Koo, Y. M. (2010). Optimization of Lipase-Catalyzed Glucose Ester Synthesis in Ionic Liquids. *Bioprocess and Biosystems Engineering*, *33*, 63-70. doi:10.1007/s00449-009-0363-4.

Hayes, D. G. (2011). Lipid-Bioprocessing Methods to Prepare Biobased Surfactants for Pharmaceutical Products. *American Pharmaceutical Review*, *14(3):* 8. doi:10.1021/acssuschemeng.2c01727.

Hollenbach, R., Delavault, A., Gebhardt, L., Soergel, H., Muhle-Goll, C., Ochsenreither, K., & Syldatk, C. (2022). Lipase-Mediated Mechanoenzymatic Synthesis of Sugar Esters in Dissolved Unconventional and Neat Reaction Systems. *ACS Sustainable*

Chemistry & Engineering, *10*(31): 10192-10202. doi: 10.1021/acssuschemeng. 2c01727.

Hollenbach, R., Ochsenreither, K. & Syldatk, C. (2020). Enzymatic Synthesis of Glucose Monodecanoate in a Hydrophobic Deep Eutectic Solvent. *International Journal of Molecular Sciences, 21(12):* 4342. doi:10.3390/ijms21124342.

Hudson, E. P., Eppler, R. K., & Clark, D. S. (2005). Biocatalysis in Semi-Aqueous and Nearly Anhydrous Conditions. *Current Opinion in Biotechnology, 16(6):* 637-643. doi:10.1016/j.copbio.2005.10.004.

Hult, K., & Berglund, P. (2007). Enzyme Promiscuity: Mechanism and Applications. *Trends in biotechnology, 25(5):* 231-238. doi:10.1016/j.tibtech.2007.03.002.

Husband, F. A., Sarney, D. B., Barnard, M. J., & Wilde, P. J. (1998). Comparison of Foaming and Interfacial Properties of Pure Sucrose Monolaurates, Dilaurate and Commercial Preparations. *Food Hydrocolloids, 12(2):* 237-244. doi:10.1016/s0268-005x(98)00036-8.

Inprakhon, P., Wongthongdee, N., Amornsakchai, T., Pongtharankul, T., Sunintaboon, P., Wiemann, L. O., Durand, A., & Sieber, V. (2017). Lipase-Catalyzed Synthesis of Sucrose Monoester: Increased Productivity by Combining Enzyme Pretreatment and Non-Aqueous Biphasic Medium. *Journal of Biotechnology, 259:* 182-190. doi:10.1016/j.jbiotec.2017.07.021.

Kapoor, M., & Gupta, M. N. (2012). Obtaining Monoglycerides by Esterification of Glycerol with Palmitic Acid Using Some High Activity Preparations of *Candida antarctica* Lipase B. *Process Biochemistry, 47*(3): 503-508. doi:10.1016/j.procbio.2011.12.009.

Khan, N. R., & Rathod, V. K. (2018). Microwave Assisted Enzymatic Synthesis of Speciality Esters: A Mini-Review. *Process Biochemistry, 75:* 89-98. doi:10.1016/j.procbio.2018.08.019.

Khan, N. R.; Rathod, V. K. (2015). Enzyme Catalyzed Synthesis of Cosmetic Esters and Its Intensification: A Review. *Process Biochemistry,* 50: 1793- 1806. doi:10.1016/j.procbio.2015.07.014.

Kim, M. J., Choi, M. Y., Lee, J. K., & Ahn, Y. (2003). Enzymatic Selective Acylation of Glycosides in Ionic Liquids: Significantly Enhanced Reactivity and Regioselectivity. *Journal of Molecular Catalysis B: Enzymatic, 26(3-6):* 115-118. doi:10.1016/j.molcatb.2003.04.001.

Kitamoto, D., Yanagishita, H., Shinbo, T., Nakane, T., Kamisawa, C., & Nakahara, T. (1993). Surface Active Properties and Antimicrobial Activities of Mannosylerythritol Lipids as Biosurfactants Produced by *Candida antarctica. Journal of Biotechnology, 29(1-2):* 91-96. doi:10.1016/0168-1656(93)90042-l.

Kralova, I., & Sjöblom, J. (2009). Surfactants Used in Food Industry: A Review. *Journal of Dispersion Science and Technology, 30(9):* 1363-1383. doi:10.1080/01932 690902735561.

Kreiner, M., Moore, B. D., & Parker, M. C. (2001). Enzyme-Coated Micro-Crystals: A 1-Step Method for High Activity Biocatalyst Preparation. *Chemical Communications, (12):* 1096-1097. doi:10.1039/b100722j.

Kulakovskaya, T. V., Kulakovskaya, E. V., & Golubev, W. I. (2003). ATP Leakage from Yeast Cells Treated by Extracellular Glycolipids of *Pseudozyma fusiformata*. *FEMS Yeast Research, 3(4):* 401-404. doi:10.1016/s1567-1356(02)00202-7.

Kumar, C. G., Mamidyala, S. K., Das, B., Sridhar, B., Devi, G. S., & Karuna, M. S. (2010). Synthesis of Biosurfactant-Based Silver Nanoparticles with Purified Rhamnolipids Isolated from *Pseudomonas aeruginosa* BS-161R. *Journal of Microbiology and Biotechnology, 20(7):* 1061-1068. doi:10.4014/jmb.1001.01018.

Laane, C., Boeren, S., Vos, K., & Veeger, C. (1987). Rules for optimization of Biocatalysis in Organic Solvents. *Biotechnology and Bioengineering, 30(1):* 81-87. doi:10.1002/bit.260300112.

Lee, S. H., Dang, D. T., Ha, S. H., Chang, W. J., & Koo, Y. M. (2008). Lipase-Catalyzed Synthesis of Fatty Acid Sugar Ester Using Extremely Supersaturated Sugar Solution in Ionic Liquids. *Biotechnology and Bioengineering, 99(1):* 1-8. doi:10.1002/bit.21534.

Li, H., Guo, W., Ma, X. J., Li, J. S., & Song, X. (2017). In Vitro and In Vivo Anticancer Activity of Sophorolipids to Human Cervical Cancer. *Applied Biochemistry and Biotechnology, 181,* 1372-1387. doi:10.1007/s12010-016-2290-6.

Li, L., Ji, F., Wang, J., Jiang, B., Li, Y., & Bao, Y. (2015). Efficient Mono-Acylation of Fructose by Lipase-Catalyzed Esterification in Ionic Liquid Co-Solvents. *Carbohydrate Research, 416,* 51-58. doi:10.1016/j.carres.2015.08.009.

Liang, M. Y., Chen, Y., Banwell, M. G., Wang, Y., & Lan, P. (2018). Enzymatic Preparation of A Homologous Series of Long-Chain 6-O-Acylglucose Esters and Their Evaluation As Emulsifiers. *Journal of Agricultural and Food Chemistry, 66(15):* 3949-3956. doi:10.1021/acs.jafc.8b00913.

Lin, X. S., Wen, Q., Huang, Z. L., Cai, Y. Z., Halling, P. J., & Yang, Z. (2015). Impacts of Ionic Liquids on Enzymatic Synthesis of Glucose Laurate and Optimization with Superior Productivity by Response Surface Methodology. *Process Biochemistry, 50(11):* 1852-1858. doi:10.1016/j.procbio.2015.07.019.

Liu, K., Sun, Y., Cao, M., Wang, J., Lu, J. R., & Xu, H. (2020). Rational Design, Properties, and Applications of Biosurfactants: A Short Review of Recent Advances. *Current Opinion in Colloid & Interface Science, 45,* 57-67. doi:10.1016/j.cocis.2019.12.005.

Liu, Q., Janssen, M. H., van Rantwijk, F., & Sheldon, R. A. (2005). Room-Temperature Ionic Liquids That Dissolve Carbohydrates in High Concentrations. *Green Chemistry, 7(1):* 39-42. doi:10.1039/b412848f.

Ljunger, G., Adlercreutz, P., & Mattiasson, B. (1994). Lipase Catalyzed Acylation of Glucose. *Biotechnology letters, 16:* 1167-1172. doi: 10.1007/BF01020845.

López, C., Cruz-Izquierdo, Á., Picó, E. A., García-Bárcena, T., Villarroel, N., Llama, M. J., & Serra, J. L. (2014). Magnetic Biocatalysts and Their Uses to Obtain Biodiesel and Biosurfactants. *Frontiers in Chemistry, 2:* 72. doi:10.3389%2ffchem.2014.00072.

Lu, X., Luo, Z., Yu, S., & Fu, X. (2012). Lipase-Catalyzed Synthesis of Starch Palmitate in Mixed Ionic Liquids. *Journal of Agricultural and Food Chemistry, 60(36):* 9273-9279. doi:10.1021/jf303096c.

Mai, N. L., Ahn, K., Bae, S. W., Shin, D. W., Morya, V. K., & Koo, Y. M. (2014). Ionic Liquids as Novel Solvents for The Synthesis of SFAE. *Biotechnology Journal, 9(12):* 1565-1572. doi:10.1002/biot.201400099.

Majumder, A. B., Singh, B., Dutta, D., Sadhukhan, S., & Gupta, M. N. (2006). Lipase Catalyzed Synthesis of Benzyl Acetate in Solvent-Free Medium Using Vinyl Acetate as Acyl Donor. *Bioorganic & Medicinal Chemistry Letters*, *16*(15): 4041-4044. doi:10.1016/j.bmcl.2006.05.006.

Mani, P., Dineshkumar, G., Jayaseelan, T., Deepalakshmi, K., Ganesh Kumar, C., & Senthil Balan, S. (2016). Antimicrobial Activities of a Promising Glycolipid Biosurfactant from A Novel Marine *Staphylococcus saprophyticus* SBPS 15. *3 Biotech, 6:* 1-9. doi:10.1007%2fs13205-016-0478-7.

Marcelino, P. R. F., Gonçalves, F., Jimenez, I. M., Carneiro, B. C., Santos, B. B., & da Silva, S. S. (2020). Sustainable Production of Biosurfactants and Their Applications. *Lignocellulosic Biorefining Technologies,* 159-183. doi:10.1002/9781119568858.ch8.

Markande, A. R., Patel, D., & Varjani, S. (2021). A Review on Biosurfactants: Properties, Applications and Current Developments. *Bioresource Technology, 330*: 124963. doi:10.1016/j.biortech.2021.124963.

Martinez-Garcia, M., Dejonghe, W., Cauwenberghs, L., Maesen, M., Vanbroekhoven, K., & Satyawali, Y. (2021). Enzymatic Synthesis of Glucose-and Xylose Laurate Esters Using Different Acyl Donors, Higher Substrate Concentrations, and Membrane Assisted Solvent Recovery. *European Journal of Lipid Science and Technology, 123*(2): 2000225. doi:10.1002/ejlt.202000225.

Maťátková, O., Kolouchová, I., Lokočová, K., Michailidu, J., Jaroš, P., Kulišová, M., Rezanka, T., & Masák, J. (2021). Rhamnolipids as a Tool for Eradication of *Trichosporon cutaneum* Biofilm. *Biomolecules, 11(11):* 1727. doi:10.3390%2fbiom11111727.

Meincken, M., Holroyd, D. L., & Rautenbach, M. (2005). Atomic Force Microscopy Study of the Effect of Antimicrobial Peptides on the Cell Envelope of *Escherichia coli*. *Antimicrobial agents and chemotherapy, 49(10):* 4085-4092. doi:10.1128%2faac.49.10.4085-4092.2005.

Miceli, R. T., Corr, D. T., Barroso, M., Dogra, N., & Gross, R. A. (2022). Sophorolipids: Anti-cancer Activities and Mechanisms. *Bioorganic & Medicinal Chemistry, 65:* 116787. doi:10.1016/j.bmc.2022.116787.

Mimee, B., Labbé, C., Pelletier, R., & Bélanger, R. R. (2005). Antifungal Activity of Flocculosin, A Novel Glycolipid Isolated from *Pseudozyma flocculosa*. *Antimicrobial Agents and Chemotherapy, 49(4):*1597-1599. doi:10.1128%2faac.49.4.1597-1599.2005.

Mimee, B., Pelletier, R., & Bélanger, R. R. (2009). In Vitro Antibacterial Activity and Antifungal Mode of Action of Flocculosin, A Membrane-Active Cellobiose Lipid. *Journal of Applied Microbiology, 107(3):*989-996. doi:10.1111/j.1365-2672.2009.04280.x.

Mishra, N., Rana, K., Seelam, S. D., Kumar, R., Pandey, V., Salimath, B. P., & Agsar, D. (2021). Characterization and Cytotoxicity of *Pseudomonas* Mediated Rhamnolipids Against Breast Cancer MDA-MB-231 Cell Line. *Frontiers in Bioengineering and Biotechnology, 9,* 761266. doi:10.3389/fbioe.2021.761266.

Moye, C. J. (1972). Non-Aqueous Solvents for Carbohydrates. In: *Advances in carbohydrate chemistry and biochemistry* (pp. 85-125). Academic Press. doi:10.1016/S0065-2318(08)60398-4.

Müller, F., Hönzke, S., Luthardt, W. O., Wong, E. L., Unbehauen, M., Bauer, J., Haag, R., Hedtrich, S., Ruhl, E. & Rademann, J. (2017). Rhamnolipids form Drug-loaded Nanoparticles for Dermal Drug Delivery. *European Journal of Pharmaceutics and Biopharmaceutics, 116:* 31-37. doi:10.1016/j.ejpb.2016.12.013.

Mulligan, C. N. (2005). Environmental Applications for Biosurfactants. *Environmental pollution, 133(2):* 183-198. doi:10.1016/j.envpol.2004.06.009.

Nagtode, V. S., Cardoza, C., Yasin, H. K. A., Mali, S. N., Tambe, S. M., Roy, P., Singh, K., Goel, A., Amin, P. D., Thorat, B. R., Cruz, J. N., & Pratap, A. P. (2023). Green Surfactants (Biosurfactants): A Petroleum-Free Substitute for Sustainability— Comparison, Applications, Market, and Future Prospects. *ACS Omega, 8(13):* 11674-11699. doi:10.1021/acsomega.3c00591.

Naughton, P. J., Marchant, R., Naughton, V., & Banat, I. M. (2019). Microbial Biosurfactants: Current Trends and Applications in Agricultural and Biomedical Industries. *Journal of Applied Microbiology, 127(1):* 12-28. doi:10.1111/jam.14243.

Neta, N. S., Teixeira, J. A., & Rodrigues, L. R. (2015). Sugar Ester Surfactants: Enzymatic Synthesis and Applications in Food Industry. *Critical Reviews in Food Science and Nutrition, 55*(5): 595-610. doi:10.1080/10408398.2012.667461.

Neu, T. R. (1996). Significance of Bacterial Surface-Active Compounds in Interaction of Bacteria with Interfaces. *Microbiological Reviews, 60(1):* 151-166. doi:10.1128%2fmr.60.1.151-166.1996.

Nguyen, P. C., Nguyen, M. T. T., Lee, C. K., Oh, I. N., Kim, J. H., Hong, S. T., & Park, J. T. (2019). Enzymatic Synthesis and Characterization of Maltoheptaose-Based Sugar Esters. *Carbohydrate polymers, 218*: 126-135. doi:10.1016/j.carbpol.2019.04.079.

Nieto, S., Villa, R., Donaire, A., & Lozano, P. (2021). Ultrasound-Assisted Enzymatic Synthesis of Xylitol Fatty Acid Esters in Solvent-Free Conditions. *Ultrasonics Sonochemistry, 75:* 105606. doi:10.1016/j.ultsonch.2021.105606.

Nitschke, M., Costa, S. G., & Contiero, J. (2010). Structure and Applications of a Rhamnolipid Surfactant Produced in Soybean Oil Waste. *Applied Biochemistry and Biotechnology, 160*: 2066-2074. doi:10.1007/s12010-009-8707-8.

Ogawa, S., Endo, A., Kitahara, N., Yamagishi, T., Aoyagi, S., & Hara, S. (2019). Factors Determining the Reaction Temperature of the Solvent-Free Enzymatic Synthesis of Trehalose Esters. *Carbohydrate Research, 482:* 107739. doi:10.1016/j.carres.2019.06.018.

Park, S., & Kazlauskas, R. J. (2001). Improved Preparation and Use of Room-Temperature Ionic Liquids in Lipase-Catalyzed Enantio-and Regioselective Acylations. *The Journal of Organic Chemistry, 66(25):* 8395-8401. doi:10.1021/jo015761e.

Pedersen, N. R., Wimmer, R., Emmersen, J., Degn, P., & Pedersen, L. H. (2002). Effect of Fatty Acid Chain Length on Initial Reaction Rates and Regioselectivity of Lipase-Catalysed Esterification of Disaccharides. *Carbohydrate Research, 337(13):* 1179-1184. doi:10.1016/s0008-6215(02)00112-x.

Pérez-Venegas, M., & Juaristi, E. (2021). Mechanoenzymology: State of the Art and Challenges Towards Highly Sustainable Biocatalysis. *ChemSusChem, 14(13):* 2682-2688. doi:10.1002/cssc.202100624.

Perinelli, D. R., Lucarini, S., Fagioli, L., Campana, R., Vllasaliu, D., Duranti, A., & Casettari, L. (2018). Lactose Oleate as New Biocompatible Surfactant for Pharmaceutical Applications. *European Journal of Pharmaceutics and Biopharmaceutics, 124:* 55-62. doi:10.1016/j.ejpb.2017.12.008.

Polat, T., Bazin, H. G., & Linhardt, R. J. (1997). Enzyme Catalyzed Regioselective Synthesis of Sucrose Fatty Acid Ester Surfactants. *Journal of Carbohydrate Chemistry, 16(9):* 1319-1325. doi:10.1080/07328309708005752.

Procentese, A., Johnson, E., Orr, V., Campanile, A. G., Wood, J. A., Marzocchella, A., & Rehmann, L. (2015). Deep Eutectic Solvent Pretreatment and Subsequent Saccharification of Corncob. *Bioresource Technology, 192:* 31-36. doi:10.1016/j.biortech.2015.05.053.

Raita, M., Laothanachareon, T., Champreda, V., & Laosiripojana, N. (2011). Biocatalytic Esterification of Palm Oil Fatty Acids for Biodiesel Production Using Glycine-Based Cross-Linked Protein Coated Microcrystalline Lipase. *Journal of Molecular Catalysis B: Enzymatic, 73(1-4):* 74-79. doi:10.1016/j.molcatb.2011.07.020.

Razafindralambo, H., Blecker, C., & Paquot, M. (2012). Carbohydrate-Based Surfactants: Structure-Activity Relationships. In *InTechOpen ebooks*. doi: 10.5772/34971.

Razafindralambo, H., Blecker, C., Mezdour, S., Deroanne, C., Crowet, J. M., Brasseur, & Paquot, M. (2009). Impacts of the carbonyl group location of ester bond on interfacial properties of sugar-based surfactants: Experimental and computational evidences. *The Journal of Physical Chemistry B, 113*(26): 8872-8877. doi:10.1021/jp903187f.

Rebello, S., Asok, A. K., Mundayoor, S., & Jisha, M. S. (2013). Surfactants: Chemistry, Toxicity and Remediation. *Pollutant Diseases, Remediation and Recycling*, 277-320. doi: 10.1007/s10311-014-0466-2.

Remichkova, M., Galabova, D., Roeva, I., Karpenko, E., Shulga, A., & Galabov, A. S. (2008). Anti-Herpesvirus Activities of *Pseudomonas* Sp. S-17 Rhamnolipid and Its Complex with Alginate. *Zeitschrift für Naturforschung C, 63(1-2):* 75-81. doi:10.1515/znc-2008-1-214.

Ren, K., & Lamsal, B. P. (2017). Synthesis of Some Glucose-Fatty Acid Esters by Lipase from *Candida antarctica* and Their Emulsion Functions. *Food Chemistry, 214:* 556-563. doi:10.1016/j.foodchem.2016.07.031.

Reyes-Duarte, D., López-Cortés, N., Ferrer, M., Plou, F. J., & Ballesteros, A. (2005). Parameters Affecting Productivity in The Lipase-Catalysed Synthesis of Sucrose Palmitate. *Biocatalysis and Biotransformation, 23(1):* 19-27. doi:10.1080/10242420500071763.

Rodrigues, L. R., Banat, I. M., van der Mei, H. C., Teixeira, J. A., Oliveira, R. (2006). Interference in Adhesion of Bacteria and Yeasts Isolated from Explanted Voice Prostheses to Silicone Rubber by Rhamnolipid Biosurfactants. *Journal of Applied Microbiology, 100,* 470-480. doi:10.1111/j.1365-2672.2005.02826.x.

Roy, I., & Gupta, M. N. (2004). Preparation of Highly Active A-Chymotrypsin for Catalysis in Organic Media. *Bioorganic & Medicinal Chemistry Letters, 14(9):* 2191-2193. doi:10.1016/j.bmcl.2004.02.026.

Saab-Rincon, G., Llopiz, A. & Arreola-Barroso, R. (2023). The Use of Biocatalysis for Biosurfactant Production. In: *Foundations and Frontiers in Enzymology Biosurfactants Research and Development* (pp. 265-301). Academic Press. doi:10.1016/b978-0-323-91697-4.00012-0.

Sachdev, D. P., & Cameotra, S. S. (2013). Biosurfactants in Agriculture. *Applied Microbiology and Biotechnology, 97*: 1005-1016. doi:10.1007%2fs00253-012-4641-8.

Saikia, J. P., Bharali, P., & Konwar, B. K. (2013). Possible Protection of Silver Nanoparticles Against Salt by Using Rhamnolipid. *Colloids and Surfaces B: Biointerfaces, 104:* 330-332. doi:10.1016/j.colsurfb.2012.10.069.

Sana, S., Datta, S., Biswas, D., & Sengupta, D. (2018). Assessment of Synergistic Antibacterial Activity of Combined Biosurfactants Revealed by Bacterial Cell Envelope Damage. *Biochimica et Biophysica Acta (BBA)-Biomembranes, 1860(2):* 579-585. doi:10.1016/j.bbamem.2017.09.027.

Sánchez, M., Teruel, J. A., Espuny, M. J., Marqués, A., Aranda, F. J., Manresa, Á., & Ortiz, A. (2006). Modulation of the Physical Properties of Dielaidoylphosphatidylethanolamine Membranes by a Dirhamnolipid Biosurfactant Produced by *Pseudomonas aeruginosa. Chemistry and Physics of Lipids, 142(1-2):* 118-127. doi:10.1016/j.chemphyslip.2006.04.001.

Santos, D. K. F., Rufino, R. D., Luna, J. M., Santos, V. A., & Sarubbo, L. A. (2016). Biosurfactants: Multifunctional Biomolecules of the 21st Century. *International Journal of Molecular Sciences, 17(3):* 401. doi:10.3390/ijms17030401.

Sarubbo, L. A., Maria da Gloria, C. S., Durval, I. J. B., Bezerra, K. G. O., Ribeiro, B. G., Silva, I. A., Twigg, M. S. & Banat, I. M. (2022). Biosurfactants: Production, Properties, Applications, Trends, and General Perspectives. *Biochemical Engineering Journal, 181:* 108377. doi:10.1016/j.bej.2022.108377.

Sarubbo, L. A., Rocha Jr, R. B., Luna, J. M., Rufino, R. D., Santos, V. A., & Banat, I. M. (2015). Some Aspects of Heavy Metals Contamination Remediation and Role of Biosurfactants. *Chemistry and Ecology, 31(8):* 707-723. doi:10.1080/02757540.2015.1095293.

Schoevaart, R., Wolbers, M. W., Golubovic, M., Ottens, M., Kieboom, A. P. G., Van Rantwijk, F., van der Wielen, L. A. M. & Sheldon, R. A. (2004). Preparation, Optimization, and Structures of Cross-Linked Enzyme Aggregates (CLEAs). *Biotechnology and Bioengineering, 87(6):* 754-762. doi:10.1002/bit.20184.

Sebatini, A. M., Jain, M., Radha, P., Kiruthika, S., & Tamilarasan, K. (2016). Immobilized Lipase Catalyzing Glucose Stearate Synthesis and Their Surfactant Properties Analysis. *3 Biotech, 6:* 1-8. doi:10.1007/s13205-016-0501-z.

Semkova, S., Antov, G., Iliev, I., Tsoneva, I., Lefterov, P., Christova, N., Nacheva, L., Stoineva, I., Kabaivanova, L. & Nikolova, B. (2021). Rhamnolipid Biosurfactants—Possible Natural Anticancer Agents and Autophagy Inhibitors. *Separations, 8(7):* 92. doi:10.3390/separations8070092.

Shah, S., Sharma, A., & Gupta, M. N. (2008). Cross-Linked Protein-Coated Microcrystals as Biocatalysts in Non-Aqueous Solvents. *Biocatalysis and Biotransformation, 26(4):* 266-271. doi:10.1080/10242420801897429.

Shah, V., Doncel, G. F., Seyoum, T., Eaton, K. M., Zalenskaya, I., Hagver, R., Azim, A. & Gross, R. (2005). Sophorolipids, Microbial Glycolipids with Anti-Human Immunodeficiency Virus and Sperm-Immobilizing Activities. *Antimicrobial Agents and Chemotherapy, 49(10):* 4093-4100. doi:10.1128%2faac.49.10.4093-4100.2005.

Shao, S. Y., Shi, Y. G., Wu, Y., Bian, L. Q., Zhu, Y. J., Huang, X. Y., Pan, Y., Zeng, L.-Y. & Zhang, R. R. (2018). Lipase-Catalyzed Synthesis of Sucrose Monolaurate and Its Antibacterial Property and Mode of Action Against Four Pathogenic Bacteria. *Molecules, 23(5):* 1118. doi:10.3390/molecules23051118.

Sheldon, R. A. (2007). Cross-Linked Enzyme Aggregates (CLEA® s): Stable and Recyclable Biocatalysts. *Biochemical Society Transactions, 35(6):* 1583-1587. doi:10.1080/10242420500183378.

Sheldon, R. A., & Woodley, J. M. (2018). Role of Biocatalysis in Sustainable Chemistry. *Chemical Reviews, 118(2):* 801-838. doi:10.1021/acs.chemrev.7b00203.

Sheldon, R. A., Lau, R. M., Sorgedrager, M. J., van Rantwijk, F., & Seddon, K. R. (2002). Biocatalysis in Ionic Liquids. *Green Chemistry, 4(2):* 147-151. doi:10.1021/cr050946x.

Shin, D. W., Mai, N. L., Bae, S. W., & Koo, Y. M. (2019). Enhanced Lipase-Catalyzed Synthesis of sugar Fatty Acid Esters Using Supersaturated Sugar Solution in Ionic Liquids. *Enzyme and Microbial Technology, 126:* 18-23. doi:10.1016/j.enzmictec.2019.03.004.

Šibalić, D., Šalić, A., Zelić, B., Tran, N. N., Hessel, V., Nigam, K. D., & Tišma, M. (2023). Synergism of Ionic Liquids and Lipases for Lignocellulosic Biomass Valorization. *Chemical Engineering Journal, 461,* 142011. doi:10.1016/j.cej.2023.142011.

Siebenhaller, S., Gentes, J., Infantes, A., Muhle-Goll, C., Kirschhöfer, F., Brenner-Weiß, G., Ochsenreither, K., & Syldatk, C. (2018). Lipase-Catalyzed Synthesis of Sugar Esters in Honey and Agave Syrup. *Frontiers in Chemistry, 6:* 24. doi:10.3389%2ffchem.2018.00024.

Siebenhaller, S., Hajek, T., Muhle-Goll, C., Himmelsbach, M., Luy, B., Kirschhöfer, F., Brenner-Weiß, G., Hahn, T., & Syldatk, C. (2017). Beechwood Carbohydrates for Enzymatic Synthesis of Sustainable Glycolipids. *Bioresources and Bioprocessing, 4(1).* doi:10.1186/s40643-017-0155-7.

Singh, P. B., & Saini, H. S. (2013). Exploitation of Agro-Industrial Wastes to Produce Low-Cost Microbial Surfactants. In: *Biotransformation of Waste Biomass into High Value Biochemicals* (pp. 445-471). New York, NY: Springer New york. doi: 10.1007/978-1-4614-8005-1_18.

Singh, R., Glick, B. R., & Rathore, D. (2018). Biosurfactants as a Biological Tool to Increase Micronutrient Availability in Soil: A Review. *Pedosphere, 28(2):* 170-189. doi:10.1016/s1002-0160(18)60018-9.

Smith, E. L., Abbott, A. P., & Ryder, K. S. (2014). Deep Eutectic Solvents (Dess) and Their Applications. *Chemical Reviews, 114(21):* 11060-11082. doi:10.1021/cr300162p.

Sobrinho, H. B. S., Luna, J. M., Rufino, R. D., Porto, A. L. F., Sarubbo, L. A. (2014). Biosurfactants: Classification, Properties and Environmental Applications. In: *Recent Developments in Biotechnology* (pp. 303-330). Stadium Press LLC, USA.

Solanki, K., & Gupta, M. N. (2008). Optimising Biocatalyst Design for Obtaining High Transesterification Activity by A-Chymotrypsin in Non-Aqueous Media. *Chemistry Central Journal, 2:* 1-7. doi:10.1186%2f1752-153x-2-2.

Solanki, K., & Gupta, M. N. (2011). A Chemically Modified Lipase Preparation for Catalyzing the Transesterification Reaction in Even Highly Polar Organic Solvents. *Bioorganic & Medicinal Chemistry Letters, 21(10)*: 2934-2936. doi:10.1016/j.bmcl.2011.03.059.

Stubenrauch, C. (2001). Sugar Surfactants—Aggregation, Interfacial, and Adsorption Phenomena. *Current Opinion in Colloid & Interface Science, 6(2):* 160-170. doi:10.1016/s1359-0294(01)00080-2.

Szűts, A., Budai-Szűcs, M., Erős, I., Otomo, N., & Szabó-Révész, P. (2010). Study of Gel-Forming Properties of Sucrose Esters for Thermosensitive Drug Delivery Systems. *International Journal of Pharmaceutics, 383(1-2):* 132-137. doi:10.1016/j.ijpharm.2009.09.013.

Thakur, P., Saini, N. K., Thakur, V. K., Gupta, V. K., Saini, R. V., & Saini, A. K. (2021). Rhamnolipid the Glycolipid Biosurfactant: Emerging trends and promising strategies in the field of biotechnology and biomedicine. *Microbial Cell Factories, 20*, 1-15. doi: 10.1186/s12934-020-01497-9.

Therisod, M., & Klibanov, A. M. (1987). Regioselective Acylation of Secondary Hydroxyl Groups in Sugars Catalyzed by Lipases in Organic Solvents. *Journal of the American Chemical Society, 109(13):* 3977-3981. doi:10.1021/ja00247a024.

Twigg, M. S., Adu, S. A., Sugiyama, S., Marchant, R., & Banat, I. M. (2022). Mono-Rhamnolipid Biosurfactants Synthesized by *Pseudomonas aeruginosa* Detrimentally Affect Colorectal Cancer Cells. *Pharmaceutics, 14(12):* 2799. doi:10.3390%2fpharmaceutics14122799.

Udomrati, S., & Gohtani, S. (2014). Enzymatic Esterification of Tapioca Maltodextrin Fatty Acid Ester. *Carbohydrate polymers, 99*, 379-384. doi:10.1016/j.carbpol.2013.07.081.

van Kempen, S. E., Schols, H. A., van der Linden, E., & Sagis, L. M. (2014). Effect of Variations in The Fatty Acid Chain on Functional Properties of Oligofructose Fatty Acid Esters. *Food Hydrocolloids, 40:* 22-29. doi:10.1016/j.foodhyd.2014.01.031.

van Rantwijk, F., Secundo, F., & Sheldon, R. A. (2006). Structure and Activity of *Candida antarctica* Lipase B in Ionic Liquids. *Green Chemistry, 8(3):* 282-286. doi:10.1039/b513062j.

Vekariya, R. L. (2017). A Review of Ionic Liquids: Applications Towards Catalytic Organic Transformations. *Journal of Molecular Liquids, 227*: 44-60. doi:10.1016/j.molliq.2016.11.123.

Vuillemin, M. E., Husson, E., Laclef, S., Jamali, A., Lambertyn, V., Pilard, S., Cailleu, D.& Sarazin, C. (2022). Improving the Environmental Compatibility of Enzymatic Synthesis of Sugar-Based Surfactants Using Green Reaction Media. *Process Biochemistry, 117*, 30-41. doi:10.1016/j.procbio.2022.03.015.

White, D. A., Hird, L. C., & Ali, S. T. (2013). Production and Characterization of a Trehalolipid Biosurfactant Produced by The Novel Marine Bacterium *Rhodococcus* Sp., Strain PML026. *Journal of Applied Microbiology, 115*(3): 744-755. doi:10.1111/jam.12287.

Yang, X. E., Zheng, P., Ni, Y., & Sun, Z. (2012). Highly Efficient Biosynthesis of Sucrose-6-Acetate with Cross-Linked Aggregates of Lipozyme TL 100 L. *Journal of Biotechnology, 161(1):* 27-33. doi:10.1016/j.jbiotec.2012.05.014.

Yi, G., Son, J., Yoo, J., Park, C., & Koo, H. (2019). Rhamnolipid Nanoparticles for In Vivo Drug Delivery and Photodynamic Therapy. *Nanomedicine: Nanotechnology, Biology and Medicine, 19:* 12-21. doi:10.1016/j.nano.2019.03.015.

Younes, M., Aquilina, G., Castle, L., Degen, G. H., Engel, K., Fowler, P., Fernández, M. J. F., Fürst, P., Gürtler, R., Husøy, T., Manco, M., Mennes, W., Moldéus, P., Passamonti, S., Shah, R., Waalkens-Berendsen, I., Wright, M., Cheyns, K., Dusemund, B., Mirat, M., Mortensen, A., Turck, D., Wölfle, D., Barmaz, S., Mech, A., Rincon, A. M., Tard, A., Vianello, G., Zakidou, P., & Gundert-Remy, U. (2023). Re-evaluation of sucrose esters of fatty acids (E 473) as a food additive in foods for infants below 16 weeks of age and follow-up of its previous evaluations as food additive for uses in foods for all population groups. *EFSA Journal, 21(4).* doi:10.2903/j.efsa.2023.7961.

Zago, E., Joly, N., Chaveriat, L., Lequart, V., & Martin, P. (2021). Enzymatic Synthesis of Amphiphilic Carbohydrate Esters: Influence of Physiochemical and Biochemical Parameters. *Biotechnology Reports, 30*: e00631. doi:10.1016/j.btre.2021.e00631.

Zaks, A., & Klibanov, A. M. (1988). Enzymatic Catalysis in Nonaqueous Solvents. *Journal of Biological Chemistry, 263(7):* 3194-3201. doi:10.1016/s0021-9258(18)69054-4.

Zhang, X., Song, F., Taxipalati, M., Wei, W., & Feng, F. (2014). Comparative Study of Surface-Active Properties and Antimicrobial Activities of Disaccharide Monoesters. *PLoS One, 9(12):* e114845. doi:10.1371%2fjournal.pone.0114845.

Zhao, G., Wang, F., Lang, X., He, B., Li, J., & Li, X. (2017). Facile One-Pot Fabrication of Cellulose Nanocrystals and Enzymatic Synthesis of Its Esterified Derivative in Mixed Ionic Liquids. *RSC Advances, 7(43):* 27017-27023. doi:10.1039/C7RA02570J.

Zheng, Y., Zheng, M., Ma, Z., Xin, B., Guo, R., & Xu, X. (2015). SFAEs, in Polar Lipids: Biology, Chemistry, and Technology, eds Ahmad M. U., & Xu X. (Urbana, IL: AOCS Press): 215–243. doi:10.1016/b978-1-63067-044-3.50012-1.

Zhu, J. P., Liang, M. Y., Ma, Y. R., White, L. V., Banwell, M. G., Teng, Y., & Lan, P. (2022). Enzymatic Synthesis of an Homologous Series of Long-And Very Long-Chain Sucrose Esters and Evaluation of Their Emulsifying and Biological Properties. *Food Hydrocolloids, 124*: 107149. doi:10.1021/acs.jafc.8b00913.

Chapter 5

Lipases: Guardians of Energy Metabolism and Beyond - Unveiling Therapeutic, Physiological, and Functional Frontiers

**Gagandeep Kaur[1],*
and Vikas Sharma[2]**

[1]Chitkara School of Pharmacy, Chitkara University, Himachal Pradesh, India
[2]Guru Gobind Singh College of Pharmacy, Yamuna Nagar, Haryana, India

Abstract

During rest and vigorous activity, fat breakdown and triacylglycerols present in skeletal muscle are crucial energy sources. Triglycerides can be broken down into their component fatty acids which are monoacylglycerols (MAGs) and diacylglycerols (DAGs) by the lipases. In addition to their importance in the digestion, absorption, and metabolism of dietary fats, lipases play a role in the digestion of a variety of lipid substrates, influencing the stability of membranes, lipid signalling, and the creation and functioning of lipid rafts. The potential for using lipases in the development of new therapeutic and diagnostic tools has grown in recent years. The multipurpose enzymes known as lipases are typically synthesized by larger organisms and are utilized to break down dietary fats and oils. Lipases play crucial roles in numerous industries, including those related to housekeeping, food preparation, medicine, and chemicals. Enzymes called lipases regulate nearly every cellular function because of their central role in lipid biochemistry. A high-fat diet, which also lowers antioxidant defences and lipoprotein lipase activity, causes abnormally high levels of lipid peroxidation, total

---

* Corresponding Author's Email: gagan17986@gmail.com.

In: Lipases and their Role in Health and Disease
Editors: Vasudeo Zambare and Mohd. Fadhil Md. Din
ISBN: 979-8-89113-628-1
© 2024 Nova Science Publishers, Inc.

cholesterol, triacylglycerol, and low-density lipoprotein in serum, as well as a decline in the level of high-density lipoprotein or cholesterol. The therapeutic applications, physiological, and functional roles of lipases will be discussed in this chapter.

Keywords: lipases, triglycerides, fatty acid, low-density lipoprotein, high fat diet, enzymes, cholesterol

Introduction

It has been known for a long time that lipids can act as signalling molecules to initiate significant physiological responses. While the vasodilatory effects of prostaglandins were being described around in the 1930s, the slow-reacting substance of anaphylaxis was also being identified (Wymann and Schneiter, 2008) Enzymes, are proteins with the ability to catalyze a wide range of chemical and biological processes. High conversion rates can be achieved with little interference from environmental factors like temperature, pressure, and pH by using these naturally occurring catalysts. In 1856, Claude Bernard found the enzyme lipase in pancreatic juice; which hydrolyzed oil droplets, turning them into more soluble byproducts (Huang et al., 2019) Lipases, also known as triacylglycerol acyl hydrolase, are a type of hydrolase that break down carboxylic ester bonds. They are members of the family of serine hydrolases and function independently of any other cofactor. Natural lipases are used to hydrolyse triglycerides into mono- and diglycerides, fatty acids, and glycerol (Kapoor and Gupta, 2012; Beisson et al., 2000). Lipases are multipurpose enzymes that have drawn interest from a wide range of industries. Animals, plants, and microorganisms are all potential resources for obtaining lipase. The pancreas and the digestive system of animals both produce lipases. Recent years have seen the release of data on the mechanistic properties of lipases. The structures of human pancreatic lipase (HPL) corroborated the hypothesis that the catalytic triad of lipases consists of Ser, Asp, and His. Traditional definitions of lipases emphasize their strong preference for polar, water-insoluble ester substrates, despite the fact that lipases are water-soluble ester hydrolases. Cholesterol esterases are a subfamily of this enzyme family. In contrast to phospholipases, which hydrolyze the acyl ester bonds of highly amphipathic phospholipids, lipases and cholesterol esterases catalyze the hydrolysis of sn-2 and sn-3 fatty acid bonds. When food is plentiful, cells store excess energy as triacylglycerols or wax esters. Examples of this can be seen in the formation of oil from seeds in plants and adipose tissue in animals. Steroidogenic tissues may benefit from the cholesterol supply that is stored in cholesteryl esters. According to research, Lipase is expressed and active in a wide variety of tissues, including the liver, adipose tissue (lipolysis), the vascular endothelial surface, the pancreas, and the small intestine (Holmes et al., 2010). Lipases available in pancreatic secretions are mainly responsible for the breakdown of fat and absorption of fat-soluble vitamins (Waldmann and Parhofer, 2019).

Lipoprotein lipase (LPL) is generated by cardiomyocytes and subsequently released. It then attaches to specific sites on the cell's outer membrane known as heparan sulphate proteoglycan (HSPG) binding sites. To facilitate the transfer of LPL into the endothelial cell (EC) lumen, a protein named glycosylphosphatidylinositol-anchored high-density lipoprotein-binding protein 1 (GPIHBP1) binds to LPL in the interstitial space and conveys it to the inner space of the vascular system. Once in the vascular lumen, LPL becomes activated and carries out its crucial role of breaking down circulating triglycerides (TG) into fatty acids. (Lee et al., 2023).

Role of Hepatic Lipase

Hepatic lipase has been studied extensively over the past decade for its role as a multifunctional protein that regulates the process of lipoprotein metabolism and conditions like atherosclerosis. Hepatic lipase, a lipolytic enzyme, promotes the hydrolysis of phospholipids and triglycerides in plasma lipoproteins. Patients with hepatic lipase deficiency have elevated levels of cholesterol and triglycerides, as well as a build-up of chylomicron remains, triglyceride-rich LDLs, high-density lipoproteins (HDLs), very low-density lipoproteins (VLDLs), etc. Higher levels of bad cholesterol and phospholipids are found in the plasma of hepatic lipase-deficient mice, just as they are in human patients. Hepatic lipase plays a significant role in determining the distribution of LDL subclasses and, consequently, atherogenic risk in humans (Santamarina-Fojo et al., 2004). There may be numerous other metabolic and genetic problems that have muddled the lipoprotein profile in certain hepatic lipase deficit individuals, although this is not always the case, this is not the case for everyone with this condition (Ruel, et al., 2003). Pancreatic lipase, an enzyme produced in the pancreas's acinar cells, plays a vital role in the digestive process. It primarily focuses its action on triacylglycerides, displaying a strong preference for them over cholesterol esters, phospholipids, and galactolipids. Besides its role in breaking down triacylglycerides, pancreatic lipase can also catalyze the hydrolysis of retinyl esters within the body.

It is evidenced by the essential function of pancreatic lipase, especially when working in conjunction with its cofactor colipase, in efficiently processing dietary fats during digestion. Given this well-established evidence, it becomes clear that modulating the activity of human pancreatic lipase holds promise for exploring new avenues in the development of therapeutic agents. These agents have the potential to inhibit fat absorption in the body, offering valuable opportunities for managing conditions such as obesity and other related metabolic disorders (Kumar and Chauhan, 2021).

Role of Pancreatic Lipase in Digestion

For optimal fat digestion and absorption, the preduodenal lipase enzyme is crucial as digestion begins in the stomach (Miller and Lowe, 2008). Lipase produced in the pancreas is essential for breaking down and absorbing lipids from food. Pancreatic triglyceride lipase is the most prominent lipolytic enzyme released. The triglyceride lipase family includes PNLIPRP2 having GenBank Accession No. HSA149D17. The amino acid sequences of PNLIPRP1 and 2 are 68 and 65% similar to that of PNLIP (Pancreatic lipase), respectively (Giller et al., 1992). Numerous experimental data support the hypothesis that lactating mammals' expression of PNLIPRP2 mRNA follows a temporal pattern indicative of a critical role in the digestion of milk fat. Curiously, the main HPNLIPRP2 enzyme was connected to the breakdown of such common vegetable lipids in the colon, yet the enzyme differs among species as an anticipated galactic lipase (Zhu et al., 2021). Probiotics in fermented milk seem to have boosted the production of a substance called fasting-related adipocyte factor. This substance acts as an inhibitor for lipoprotein lipase, which, in turn, results in a decrease in the storage of fat within adipose tissues. Furthermore, when milk undergoes fermentation with these probiotics, it produces short-chain fatty acids (SCFAs). These SCFAs possess anti-obesity characteristics as they attach to receptors for free fatty acids in the intestines. This attachment promotes the oxidation of fatty acids and effectively hinders the build-up of fat (Sakandar and Zhang, 2021).

Triacylglyceride Lipases Acts in Series to Regulate Lipolysis

Adipose triglyceride lipase (ATGL) is the primary enzyme responsible for intracellular lipolysis that yields fatty acids (FAs) for energy synthesis from TG reserves. As Abnormal TG accumulation in numerous organs and tissues, Neutral Lipid Storage Disease-Associated Myopathy (NLSD-M) is a hereditary condition that affects humans and causes cardiac and skeletal muscle myopathy. Dysregulated lipolysis is linked to metabolic illnesses such as insulin resistance, diabetes, inflammation, and non-alcoholic fatty liver. The lipase is responsible for the initial step in the breakdown of TGs. At the outset of this process, a protein known as ATGL catalyses the conversion of TG to DAG and FA. Hormone-sensitive lipase (HSL) then converts DG to monoacylglycerol (MG) and FA. Monoglyceride lipase (MGL) on the other hand, breaks down MG into glycerol and FA. Insulin resistance, type 2 diabetes, fatty liver, and inflammation are only a few of the negative outcomes of an organism with an excess of circulating FAs due to dysregulated lipolysis (Schweiger et al., 2009; Cerk et al., 2018).

Functions of Triacylglyceride Lipase

The heart, testicles and muscles of the skeleton all exhibit various degrees of ATGL expression, although brown and white adipose tissue exhibits the highest levels of ATGL expression. ATGL-null animals, as shown by functional investigations, have a larger adipose tissue mass and greater Triglyceride deposition can occur in various tissues, including adipose tissue as well as the liver, kidney, heart, and skeletal muscles and its functions are presented in Figure 1. The metabolic equilibrium within adipocytes can undergo a transition from an anabolic state characterized by elevated transacylase activity to a catabolic state characterized by increased lipase activity. The specific mechanisms that regulate the transacylation and lipase activities of ATGL in this context remain unclear. This could have an impact on the distribution of substrate to peripheral tissues. The control of HSL, which plays a crucial role as Triacylglycerols and diacylglycerols lipase, has been extensively studied. Recent studies have demonstrated the significance of ATGL and its associating partner CGI-58 in triglyceride (TAG) lipolysis and the maintenance of metabolic homeostasis across various bodily functions. Mice lacking HSL collected DAG and maintained both adrenergic-stimulated and basal lipolysis, indicating the need for additional enzymes involved in TAG hydrolysis. First, ATGL removes the first fatty acid from triacylglyerols to generate DAG, which then undergoes hydrolysis by HSL to provide an additional FA alongside the MAG substrate, thereby initiating lipolysis. In the last stage of lipolysis, monoacylglycerols lipase converts MAGs into fatty acids and glycerol (Haemmerle et al., 2006, Watt and Steinberg, 2008). TGs have essential functions in the storage of fatty acids and as building blocks for membrane lipids. These stored fatty acids are critical for generating energy and constructing cellular membranes during the process of cell division. When the body requires energy, these fatty acids are released and transported to various tissues, including muscle cells, where they undergo oxidation to produce adenosine triphosphate (ATP), which serves as the primary energy currency in the body. In the late G1 phase, lipolysis is activated, allowing TGs to be broken down by lipases. This process results in the production of fatty acids and the components needed for building cellular membranes (Nakatsukasa, et al., 2022).

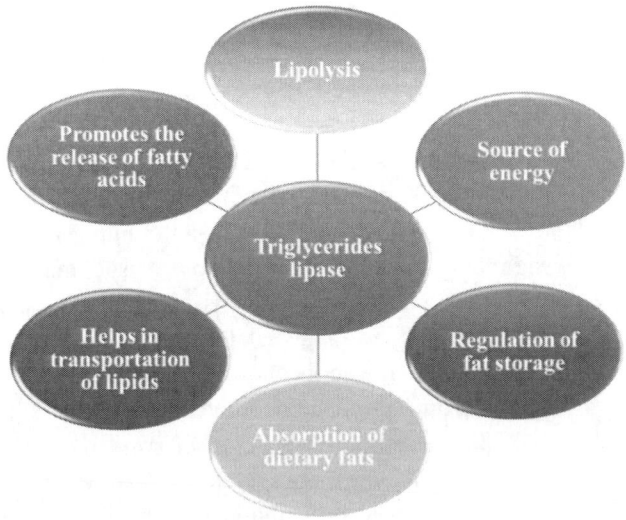

Figure 1. Functions of Triglycerides Lipase.

Effects of Inflammation on Lipid Metabolism

Both homeostasis and immunity can be modulated by inflammation and lipid signalling pathways. New research indicated that many lipid species, along with well-studied eicosanoids and inositol phospholipids, acted positively and negatively on regulating inflammatory responses. Enzymes such as phospholipase A and cyclooxygenase work in combination to change the arachidonic acid found in membrane phospholipids into eicosanoids. In response to pathogen-associated molecular patterns (PAMPs) and pro-inflammatory cytokines, two distinct classes of enzymes are directly regulated, allowing for quick and substantial increases in prostaglandin synthesis. The resulting prostaglandin metabolites perform a variety of biological roles, including escalating inflammatory responses that underlie the "cardinal signs" of inflammation. It was once thought that catecholamine release caused the hyperlipidaemia of sepsis because it prompted the discharge of free fatty acids from adipose tissue, which the liver then processed and excreted in the form of VLDL. The pace at which triglycerides are cleared from the body's periphery is reduced when TNF-α is present, which may contribute to hypertriglyceridemia (Nonogaki *et. al.*, 1994; Glass and Olefsky, 2012).

Lipases and Lipolysis in Lipid Metabolism

Lipolysis is the breakdown of glycerides found in lipid droplets within cells. Gastric lipase, Lingual lipase, and pancreatic lipases are the most important lipolytic enzymes in the gastrointestinal tract. LPL and hepatic TG lipases are required for TG hydrolysis in the vascular system. Vascular lipases are physically similar to pancreatic lipases and both serve as the prototype member. Lysosomal acid lipases (pH optimum between pH 4-5) and neutral lipases (pH optimum around pH 7) are involved in the intracellular lipolysis of TGs. The most significant lipase in lysosomes is called lysosomal acid lipase (LAL), while hormone-sensitive lipase (HSL) and ATGL are examples of well-characterized neutral TG hydrolases. Pathogenesis of cancer-associated cachexia (CAC) may also involve lipolytic signalling. Muscle and fat loss caused by tumours were prevented in ATGL-deficient mice, as shown in a recent study (Das et al., 2011). Mice lacking HSL showed some degree of protection as well. ATGL-deficient animals maintained a normal body weight and composition despite having higher levels of the circulating substances that trigger apoptosis, lipolysis, and muscle protein degradation. This implies that lipolysis contributes to the signalling network that ultimately results in the reduction of adipose tissue and skeletal muscle mass. The catabolic part of the FA cycle, lipolysis, supplies needed FAs during the metabolic needs of the body and eliminates them during excess. For example, lipids found in membranes and those that are crucial for cellular signalling are both synthesized from FAs (Unger et al., 2010). The initial recognition of the signalling capabilities of MGs stemmed from the discovery that 2-arachidonoylglycerol (2-AG), an MG generated from phospholipids, exerts its effects by activating the cannabinoid receptor (CBR). This activation of CBR by 2-AG plays a crucial role in regulating energy balance, lipid metabolism, and food intake. The endocannabinoid system consists of a set of receptors, including CB1 and CB2, which are associated with G proteins. Additionally, this system includes endocannabinoids and enzymes that play a role in the manufacture and breakdown of ECs. Exogenous fatty acids are supplied from adipose tissue, a specialized organ in higher species, liver and muscle are two more high-demand tissues FA esterification and TG hydrolysis work together to form an effective buffer system that allows for adequate FA flux without leading to non-physiological elevations in cellular concentrations of non-esterified FA (Di Marzo, 2009).

Therapeutic Role of Lipases

In humans, lipases may be involved in fat digestion. Lipases have expanded in scope and are developing quickly as potential treatment possibilities. In the pharmaceutical industry, lipases have emerged as promising "drug" candidates and "marker" enzymes, and they are also being used as diagnostic tools for identifying infectious and disease states. Colorimetric enzyme-based detection of glycerol release is used in the enzymatic measurement of serum triglycerides. The diagnosis of acute pancreatitis or pancreatic damage may be aided by measuring serum lipase levels. Multi-step processes involving the transport of different cofactors, bile salts, and catalysts are required for the breakdown of dietary fats, particularly triacylglycerides. Ester bonds connect 3'-unsaturated fatty acid chains to the glycerol backbone in triglyceride atoms (Jawed et al., 2019). Human gastric lipases, lingual lipases, etc., are just two examples of the many of lipases found in the digestive system. Colorimetric enzyme-based detection of glycerol release is used in the enzymatic measurement of serum triglycerides. The identification and determination of acute pancreatitis or pancreatic damage may be aided by measuring serum lipase levels (Lott and Lu, 1991).

Digestive Enzymes

The gastrointestinal tract synthesizes and releases digestive enzymes to break down proteins, fats, and carbohydrates, facilitating the process of digestion and subsequent absorption of essential nutrients. Supplementation, when deemed appropriate, can serve as a dependable adjunctive therapy for many conditions characterized by compromised digestive functioning (Ianiro et al., 2016). Studies have shown that the use of pancreatic enzyme supplements can be advantageous for individuals with functional dyspepsia, leading to a significant reduction in symptoms such as gas, abdominal bloating, burping, a sensation of fullness, and post-meal discomfort. The rationale behind employing digestive enzymes lies in the fact that they help break down carbohydrates, proteins, and fats into smaller components initially, which are then absorbed by the body (Swami and Shah, 2017). Alpha amylase breaks down starch and glycogen into glucose, while glucoamylase specifically hydrolyses the non-reducing terminal of starch to produce glucose. On the other hand, alpha galactosidase is responsible for breaking down glycolipids and glycoproteins. Lactase is an essential enzyme that serves a critical function in the hydrolysis of lactose, hence facilitating the thorough digestion of whole milk. Lactase demonstrates efficacy in those who have signs of lactose intolerance.

The enzyme invertase facilitates the hydrolysis process, resulting in the conversion of sucrose into fructose and glucose. Probiotics are living microorganisms, and when consumed in sufficient quantities, they promote improved gastrointestinal health. Probiotic supplements have been shown to effectively improve the overall health of persons diagnosed with functional Gastrointestinal Disorders, a category of conditions characterized by symptoms such as belly pain, dysbiosis, and dyspepsia, among others (Qamra et al., 2020).

Fat Digestion

The process of digesting dietary fats, particularly triacylglycerol, is a complex and sophisticated one that requires the participation of several cofactors, bile salts, and enzymes. Triacylglycerol molecules consist of a glycerol backbone whereby three unsaturated chains of fatty acids are connected via ester bonds. This entire absorption process initiates within the oral cavity of animals and engages several enzymes, including lipases.

As fats are broken down into fatty acids and glycerol, they can traverse the small intestine's walls and then be transported via the bloodstream for immediate energy utilization or storage for future requirements. Within the gastrointestinal system, numerous lipase variations are present, such as human gastric lipases, lingual lipases, and pancreatic lipases. Each of these lipases serves a distinct role in this complex digestive sequence (Jawed et al., 2019).

Pharmaceutical Applications

Lipases have a significant impact on pharmaceutical applications, particularly in processes like trans-esterification and hydrolysis. These enzymes are produced by various microorganisms, including bacteria, yeasts, molds, and some protozoa, to break down lipid materials. Interestingly, these microorganisms can thrive in challenging environmental conditions. Lipases derived from *C. rugosa* have demonstrated the ability to lower serum cholesterol levels. Furthermore, lipases sourced from S. marcescens have been instrumental in synthesizing lovastatin, a medication. This particular lipase played a pivotal role in the asymmetric hydrolysis of 3-phenylglycidic acid ester, a crucial intermediate in the production of diltiazem hydrochloride (Choudhury and Bhunia et al., 2015; Gopinath et al., 2023).

Treatment of Pancreatic Insufficiency

Chronic pancreatitis (CP) and cystic fibrosis (CF) are recognized as the predominant etiologies of pancreatic insufficiency, a pathological state characterized by inadequate production of digesting enzymes, specifically lipases, by the pancreas. When this occurs, lipase supplements are administered to aid patients in processing dietary fats and preventing issues linked to malabsorption, such as the presence of fatty stools (steatorrhea) and malnutrition. Diabetes can also lead to exocrine insufficiency.

In contrast, certain conditions like Celiac sprue, are linked to decreased pancreatic lipase production without causing damage to the pancreas itself. Reversible pancreatic insufficiency is sometimes observed in premature infants due to their developmental immaturity, with their pancreas typically reaching full functionality by the age of 2 years. Motility problems characterized by accelerated stomach emptying and decreased transit time in the small intestine can also lead to malabsorption as a result of insufficient lipid digestion. The heightened velocity at which food, bile, and pancreatic enzymes traverse the digestive system impedes optimal blending, diminishes the duration available for engagement with the small intestine, compromises the process of digestion and absorption, and influences the stimulation of pancreatic function (Bruno et al, 1995; Whitcomb et al., 2010; Fieker, et al., 2011).

Use in Drugs and Diagnostic Tools

Lipases have always been effectively utilised for the purpose of generating optically pure enantiomeric compounds by the separation of their dextro and levo forms. The process of resolving racemic mixtures holds significant importance in the synthesis, effectiveness, and purity of medicinal compounds, as only one isomeric form often exhibits therapeutic activity. Similar approaches have been employed in the manufacturing of many medications inside multiple pharmaceutical establishments. Lipases have been utilized in the production of various non-steroidal anti-inflammatory drugs (NSAIDs), including naproxen, ibuprofen, and lubeluzole (a medication used for stroke prevention), and diltiazem (a medication that blocks calcium channels) for the treatment of hypertension and angina (Godoy et al., 2022). The utilisation of lipases derived from *Candida rugosa* and *Serratia marcescens*, which are microorganisms, have been exploited in the synthesis of valuable pharmaceutical compounds due to their ability to catalyse enantioselective processes. The study conducted by Rafiee and Rezaee (2021) employed immobilised lipase derived from *S. marcescens* ECU1010 (Sml) on several support mediums, namely Chitosan, Celite 545, and DEAE cellulose. This immobilised lipase was utilised in a bioreactor to enhance the efficiency and performance of the enantioselective hydrolysis of (±) methyl trans-3(4-methoxyphenyl) glycidate (MPGM). Lipases have emerged as significant prospects for therapeutic interventions and as enzyme markers in the pharmaceutical industry (Chandra et al., 2020). They are employed as diagnostic instruments to assess lipase levels, which can indicate the presence of infections or certain disorders. Certain lipases are utilised for enzymatic evaluation of blood triglycerides, in which the liberation of glycerol is identified using enzyme-driven colorimetric detection (Pàmies and Bäckvall, 2003). In a study conducted by Søreide et al., (2018), it was found that the assessment of lipase concentrations in the circulatory system can be utilized as a diagnostic instrument for the identification and evaluation of acute pancreatitis. Researchers have put out a proposal suggesting that the detection of elevated levels of pancreatic lipase in the circulatory system could potentially act as a biomarker for the diagnostic purposes of acute pancreatitis. In one study of Kim et al., (2020) developed fluorescent-labelled versions of triacylglycerol and cholesterol in order to provide a fluorometric assay designed to measure LAL activity. Following this, the functionality of LAL can be linked to the identification of Wolman's disease and cholesteryl ester storage disease (CESD). Individuals with a disease display a reduced degree of LAL activity, resulting in a decline in fluorescence intensity. In the context of Wolman's disease, lipases have dual advantages, namely in terms of diagnostic utility and therapeutic efficacy. Biosensors have been devised to detect and quantify lipids, becoming them a significant diagnostic instrument. Aggarwal and Pundir, (2016) constructed an amperometric triglyceride biosensor mechanism by immobilizing glycerol lipase kinase and glycerol phosphate oxidase onto media of polyvinyl-alcohol. Devices with similar capabilities can be employed to rapidly and efficiently diagnose levels of blood/serum cholesterol and triacylglycerides, which are significant indications of several diseases.

Microbial and Plant-Based Lipase Inhibitors for Obesity Prevention

Obesity is one of the complex conditions characterized by an imbalance between weight and height, caused by excessive lipid buildup in adipose tissue. This is mostly attributed to the consumption of excessive calories and insufficient physical activity, leading to an energy imbalance. Optimising the energy that contributes towards obesity poses a challenging task. Pancreatic lipase plays a crucial role in the digestion of TAGs within the intestine, facilitating their absorption. Consequently, it can be regarded as the most essential enzyme for the breakdown of dietary fat. Therefore, the inhibition of pancreatic lipases will result in a decrease in the intestinal absorption of lipids. Tetrahydrolipstatin (THL), a compound developed from lipstatin (Rajan et al., 2020), first isolated from Streptomyces toxytricini, has gained significant recognition as a powerful inhibitor of many enzymes, including steroid ester hydrolases, pancreatic as well as gastric lipases etc. Currently, efforts are underway to commercialise THL as a pharmaceutical agent for the treatment of obesity. The amphiphilic character of THL enables its interaction with the equally significant amino acid serine located somewhere on the active site of the lipase. When bile salts are present, lipases undergo partitioning by forming micelles. This process leads to a diminished efficacy of certain amphiphilic inhibitors. Nevertheless, this is accompanied by the presence of the triacylglycerol phase, which may serve as a prerequisite for the effectiveness of lipase inhibitors, even in the presence of bile and lipids (Zhong et al., 2020).

Application of Lipases in Cancer Treatment

The correlation between a sedentary lifestyle and excessive calorie intake has been linked to an elevated susceptibility to the development of breast, pancreas, liver, colon, and prostate cancers. The amounts of TG present in the serum may have an impact on the development of colorectal and pancreatic cancers, as well as precancerous lesions. Hence, the recognition of the importance of LPL in facilitating the hydrolysis of plasma TG is also accredited (Rudrapal et al., 2022). Cachexia, a condition characterised by the loss of skeletal muscle and adipose tissue, is commonly observed in patients with different types of cancer. This debilitating condition has been found to be linked to disturbances in metabolic pathways for lipids and triglycerides. LPL exerts its influence on TGs and monoglycerides, hence assuming a critical function in lipid metabolism and lipoprotein metabolism. Cachexia triggers the modulation of LPL by several factors, such as tumour necrosis factor (TNF)-α and Interleukins (IL-1, IL-6), which subsequently restrict LPL activity. This inhibition results in a significant reduction in the deposition of fat in tissues (Nawaz et al., 2017). The phenomenon of emaciation in cachexia, which is impacted by reduced lipoprotein lipase (LPL) activity, has proven resistant to several dietary interventions. However, a study conducted by Beaudry and Law, (2022) demonstrated that the administration of an activator of LPL to a rat of cachexia and Leydig cell tumour exhibited advantageous outcomes.

Future Perspectives

Lipases have demonstrated considerable significance throughout various industrial areas, showcasing wide-ranging usefulness. Consequently, a significant body of research efforts pertaining to this enzyme encompasses various realms of knowledge. Lipases find extensive application mostly in the biotechnology and bioprocess sectors, encompassing many industries such as food, cosmetics, biofuels, and environmental domains. This observation illustrates the complex and diverse characteristics of lipase enzymes.

Numerous studies have underscored the significance of enhancing lipases for diverse objectives, encompassing their utilisation as scaffolds for immobilisation, in fermentative procedures for enzyme synthesis, enzymatic purification, biochemical characterization, and their involvement in reactions encompassing both organic and inorganic mediums. The aforementioned findings highlight the extensive array of potential uses for this particular biocatalyst. The examination of kinetics is of paramount importance in the scaling up of a process from the laboratory level to industrial production.

Lipases, also known as triacylglycerol acyl hydrolases, are enzymatic proteins that are synthesised by higher organisms. These enzymes play a crucial role in the hydrolysis of dietary oils and fats. Lipases are known to possess substantial importance in several industries, including detergent, food, pharmaceutical, and chemical sectors. The relevance of these enzymes arises from their ability to catalyse a wide range of processes, including transesterification, esterification, aminolysis, alcoholysis, and acidolysis. In recent years, there has been a broadening of the scope of their utilisation in the medical field, encompassing diverse applications such as the advancement of assistances in fat metabolism, substitute remedy, and the exploration and creation of diagnostic techniques used in the identification and assessment of metabolic abnormalities or diseases. One potential application of Lovastatin, a lipase inhibitor, is its usage as an antiobesity medication. Additionally, the "similase" enzyme, developed by Integrative Therapeutics, shows promise as a therapeutic agent for digestive problems such as sensitive stomach or pancreatic diseases. The comprehensive comprehension of lipase's modulation activities remains in an exploratory phase due to the incomplete understanding and elucidation of its mechanism of action. Porphyrin ring structures have been observed in plant and microbial lipase inhibitors which further act as effective agents in addressing metabolic diseases, including obesity. The promotion of dietary supplements, in conjunction with other preventative measures, may be advocated to mitigate the risk of lifestyle-related diseases.

This study investigates the essential role of lipases in variety of industries that hold substantial commercial value within the food and nutraceutical industries. As a result, there is an increasing need for these proteins in diverse manifestations. Enzymes are a feasible alternative method for the production of new esters. In recent years, there have been limitations in the availability and cost of lipases, which has hindered their industrial applicability. The advent of novel enzymes exhibiting improved characteristics, including heightened durability, elevated productivity, and better stability for numerous reactions, has become lipases a feasible and economically advantageous instrument in many sectors, notably the food and nutraceutical industry. The aforementioned advancement has not only facilitated the availability of lipases but also rendered them environmentally compatible, hence providing significant advantages to consumers' well-being. It is expected that future developments in the application of biocatalysts will continue to reduce the number of process stages and reaction times required for the manufacturing of new products. The domain of enzyme technology, particularly concerning lipases, comprises a diverse array of disciplines, such as protein engineering, immobilisation methodologies, process engineering, and life cycle analysis. A comprehensive understanding of the underlying mechanisms involved in lipase catalysis is crucial in order to facilitate its practical implementation with minimal efficiency losses in various processes. These areas of expertise are considered significant in this regard.

Conclusion

Triacylglycerol acyl hydrolases, sometimes referred to as lipases, are helpful and adaptable enzymes produced by more sophisticated creatures capable of metabolizing dietary oils and lipids. Hepatic lipase has a significant impact on atherogenesis because it is involved in both cellular lipid absorption and plasma lipid metabolism. Hepatic lipase has been found to have a ligand-binding activity that enhances the absorption of lipoproteins as well as lipids via receptors on cells and proteoglycans in addition to its function being a lipolytic enzyme that alters LDL and HDL. The lipase enzyme is involved in many different diseases and is crucial to the body's normal functioning. In the recent decade, much evidence has accumulated supporting lipases' role as a protein with multiple functions that control atherosclerosis and lipoprotein metabolism. Hepatic lipase performs a variety of crucial processes, including digestion and lipolysis. Triglyceride lipases act in series to regulate lipolysis, effects of inflammation on lipid metabolism and lipolysis in lipid metabolism and various other diseases as well.

References

Aggarwal, V., & Pundir, C. S. (2016). Rational Design of Nanoparticle Platforms for "Cutting-the-Fat": Covalent Immobilization of Lipase, Glycerol Kinase, and Glycerol-3-Phosphate Oxidase on Metal Nanoparticles. *Methods in Enzymology, 571:* 197-223. Academic Press. doi: 10.1016/bs.mie.2016.01.022.

Beaudry, A. G., & Law, M. L. (2022). Leucine Supplementation in Cancer Cachexia: Mechanisms and a Review of the Pre-clinical Literature. *Nutrients, 14(14):* 2824. doi: 10.3390/nu14142824.

Beisson, F., Tiss, A., Rivière, C., & Verger, R. (2000). Methods for Lipase Detection and Assay: A Critical Review. *European Journal of Lipid Science and Technology, 102(2):* 133-153. doi: 10.1002/(SICI)1438-9312(200002)102:2<133::AID-EJLT133>3.0.CO;2-X.

Bruno, M. J., Haverkort, E. B., Tytgat, G. N., & Van Leeuwen, D. J. (1995). Maldigestion Associated with Exocrine Pancreatic Insufficiency: Implications of Gastrointestinal Physiology and Properties of Enzyme Preparations for a Cause-Related and Patient-Tailored Treatment. *American Journal of Gastroenterology, 90(9):* 183-1393.

Cerk, I. K., Wechselberger, L., & Oberer, M. (2018). Adipose Triglyceride Lipase Regulation: An Overview. *Current Protein & Peptide Science, 19(2):* 221–233. doi: 10.2174/1389203718666170918160110.

Chandra, P., Enespa, Singh, R., & Arora, P. K. (2020). Microbial Lipases and Their Industrial Applications: A Comprehensive Review. *Microbial Cell Factories, 19:* 1-42. doi: 10.1186/s12934-020-01428-8.

Choudhury, P., & Bhunia, B. (2015). Industrial Application of Lipase: A Review. *Biopharma Journal, 1(2):* 41-47.

Das, S. K., Eder, S., Schauer, S., Diwoky, C., Temmel, H., Guertl, B., Gorkiewicz, G., Tamilarasan, K. P., Kumari, P., Trauner, M., Zimmermann, R., Vesely, P., Haemmerle, G., Zechner, R., & Hoefler, G. (2011). Adipose Triglyceride Lipase Contributes to Cancer-Associated Cachexia. *Science, 333(6039):* 233-238. doi: 10.1126/science.11989.

Di Marzo, V. (2009). The Endocannabinoid System: Its General Strategy of Action, Tools for its Pharmacological Manipulation and Potential Therapeutic Exploitation. *Pharmacological Research, 60(2):* 77-84. doi: 10.1016/j.phrs.2009.02.010.

Fieker, A., Philpott, J., & Armand, M. (2011). Enzyme Replacement Therapy for Pancreatic Insufficiency: Present and Future. *Clinical and Experimental Gastroenterology, 1(4):* 55-73. doi: 10.2147/CEG.S17634.

Giller, T., Buchwald, P., Blum-Kaelin, D., & Hunziker, W. (1992). Two Novel Human Pancreatic Lipase Related Proteins, hPLRP1 and hPLRP2. Differences in Colipase Dependence and in Lipase Activity. *Journal of Biological Chemistry, 267(23):* 16509-16516.

Glass, C. K., & Olefsky, J. M. (2012). Inflammation and Lipid Signalling in the Etiology of Insulin Resistance. *Cell metabolism, 15(5):* 635-645. doi: 10.1016/j.cmet.2012.04.001.

Godoy, C. A., Pardo-Tamayo, J. S., & Barbosa, O. (2022). Microbial Lipases and Their Potential in the Production of Pharmaceutical Building Blocks. *International Journal of Molecular Sciences, 23(17):* 9933. doi: 10.3390/ijms23179933.

Gopinath, S. C., Anbu, P., Lakshmipriya, T., & Hilda, A. (2013). Strategies to Characterize Fungal Lipases for Applications in Medicine and Dairy Industry. *BioMed Research International, 2013:* 154549. doi: 10.1155/2013/154549.

Haemmerle, G., Lass, A., Zimmermann, R., Gorkiewicz, G., Meyer, C., Rozman, J., Heldmaier, G., Maier, R., Theussl, C., Eder, S., Kratky, D., Wagner, E. F., Klingenspor, M., Hoefler, G., & Zechner, R. (2006). Defective Lipolysis and Altered Energy Metabolism in Mice Lacking Adipose Triglyceride Lipase. *Science, 312(5774):* 734-737. doi: 10.1126/science.112396.

Holmes, R. S., Cox, L. A., & VandeBerg, J. L. (2010). Comparative Studies of Mammalian Acid Lipases: Evidence for a New Gene Family in Mouse and Rat (Lipo). *Comparative Biochemistry and Physiology Part D: Genomics and Proteomics, 5(3):* 217-226. doi: 10.1016/j.cbd.2010.05.004.

Huang, Y., Ren, J., & Qu, X. (2019). Nanozymes: Classification, Catalytic Mechanisms, Activity Regulation, and Applications. *Chemical Reviews, 119(6):* 4357-4412. doi: 10.1021/acs.chemrev.8b00672.

Ianiro, G., Pecere, S., Giorgio, V., Gasbarrini, A., & Cammarota, G. (2016). Digestive Enzyme Supplementation in Gastrointestinal Diseases. *Current Drug Metabolism, 17(2):* 187-193. doi: 10.2174/1389200217021601141501377.

Jawed, A., Singh, G., Kohli, S., Sumera, A., Haque, S., Prasad, R., & Paul, D. (2019). Therapeutic Role of Lipases and Lipase Inhibitors Derived from Natural Resources for Remedies Against Metabolic Disorders and Lifestyle Diseases. *South African Journal of Botany, 120:* 25-32. doi: 10.1016/j.sajb.2018.04.004.

Kapoor, M., & Gupta, M. N. (2012). Lipase Promiscuity and Its Biochemical Applications. *Process Biochemistry, 47(4):* 555-569. doi: 10.1016/j.procbio.2012.01.011.

Kim, S., Kim, N., Park, S., Jeon, Y., Lee, J., Yoo, S. J., Lee, J. W., Moon, C., Yu, S. W., & Kim, E. K. (2020). Tanycytic TSPO Inhibition Induces Lipophagy to Regulate Lipid Metabolism and Improve Energy Balance. *Autophagy, 16(7):* 1200-1220. doi: 10.1080/15548627.2019.1659616.

Kumar, A., & Chauhan, S. (2021). Pancreatic lipase inhibitors: The road voyaged and successes. *Life Sciences, 271:* 119115. doi: 10.1016/j.lfs.2021.119115.

Lee, C. S., Zhai, Y., & Rodrigues, B. (2023). Changes in Lipoprotein Lipase in the Heart Following Diabetes Onset. *Engineering, 20:* 19-25. doi: 10.1016/j.eng.2022.06.013.

Lott, J. A., & Lu, C. J. (1991). Lipase Isoforms and Amylase Isoenzymes: Assays and Application in the Diagnosis of Acute Pancreatitis. *Clinical chemistry, 37(3):* 361-368.

Miller, R., & Lowe, M. E. (2008). Carboxyl Ester Lipase from Either Mother's Milk or the Pancreas is Required for Efficient Dietary Triglyceride Digestion in Suckling Mice. *The Journal of Nutrition, 138(5):* 927-930. doi: 10.1093/jn/138.5.927.

Nakatsukasa, K., Fujisawa, M., Yang, X., Kawarasaki, T., Okumura, F., & Kamura, T. (2022). Triacylglycerol Lipase Tgl4 is a Stable Protein and its Dephosphorylation is Regulated in a Cell Cycle-Dependent Manner in *Saccharomyces cerevisiae*.

Biochemical and Biophysical Research Communications, 626: 85-91. doi: 10.1016/j.bbrc.2022.08.022.

Nawaz, R., Zahid, S., Idrees, M., Rafique, S., Shahid, M., Ahad, A., Amin, I., Almas, I., & Afzal, S. (2017). HCV-Induced Regulatory Alterations of IL-1β, IL-6, TNF-α, and IFN-ϒ Operative, Leading Liver En-Route to Non-Alcoholic Steatohepatitis. *Inflammation Research, 66:* 477-486. doi: https://doi.org/10.1007/s00011-017-1029-3.

Nonogaki, K., Moser, A. H., Feingold, K. R., & Grunfeld, C. (1994). Alpha-Adrenergic Receptors Mediate the Hypertriglyceridemia Induced by Endotoxin, But Not Tumor Necrosis Factor, in Rats. *Endocrinology, 135(6):* 2644-2650. doi: 10.1210/endo.135.6.7988454.

Pàmies, O., & Bäckvall, J. E. (2003). Combination of Enzymes and Metal Catalysts. A Powerful Approach in Asymmetric Catalysis. *Chemical Reviews, 103(8):* 3247-3262. doi: 10.1021/cr020029g.

Qamra, A., Soni, N. K., Trivedi, H. H., Kumar, S., Prakash, A., Roy, S., & Mukherjee, S. (2020). A Review of Digestive Enzyme and Probiotic Supplementation for Functional Gastrointestinal Disorders. *The Indian Practitioner, 73(3),* 35-39.

Rafiee, F., & Rezaee, M. (2021). Different Strategies for the Lipase Immobilization on the Chitosan Based Supports and Their Applications. *International Journal of Biological Macromolecules, 179:* 170-195. doi: 10.1016/j.ijbiomac.2021.02.198.

Rajan, L., Palaniswamy, D., & Mohankumar, S. K. (2020). Targeting Obesity with Plant-Derived Pancreatic Lipase Inhibitors: A Comprehensive Review. *Pharmacological Research, 155:* 104681. doi: 10.1016/j.phrs.2020.104681.

Rudrapal, M., Khairnar, S. J., Khan, J., Dukhyil, A. B., Ansari, M. A., Alomary, M. N., Alshabrmi, F. M., Palai, S., Deb, P. K., & Devi, R. (2022). Dietary Polyphenols and Their Role in Oxidative Stress-Induced Human Diseases: Insights into Protective Effects, Antioxidant Potentials and Mechanism(s) of Action. *Frontiers in Pharmacology, 13:* 283. doi: 10.3389/fphar.2022.806470.

Ruel, I. L., Couture, P., Gagné, C., Deshaies, Y., Simard, J., Hegele, R. A., & Lamarche, B. (2003). Characterization of a Novel Mutation Causing Hepatic Lipase Deficiency among French Canadians. *Journal of Lipid Research, 44(8):* 1508-1514. doi: 10.1194/jlr.M200479-JLR200.

Sakandar, H. A., & Zhang, H. (2021). Trends in Probiotic (s)-Fermented Milks and Their in Vivo Functionality: A Review. *Trends in Food Science & Technology, 110:* 55-65. doi: 10.1016/j.tifs.2021.01.054.

Santamarina-Fojo, S., González-Navarro, H., Freeman, L., Wagner, E., & Nong, Z. (2004). Hepatic Lipase, Lipoprotein Metabolism, and Atherogenesis. *Arteriosclerosis, Thrombosis, and Vascular Biology, 24(10):* 1750-1754. doi: 10.1161/01.ATV.0000140818.00570.2d.

Schweiger, M., Lass, A., Zimmermann, R., Eichmann, T. O., & Zechner, R. (2009). Neutral Lipid Storage Disease: Genetic Disorders Caused by Mutations in Adipose Triglyceride Lipase/PNPLA2 or CGI-58/ABHD5. *American Journal of Physiology-Endocrinology and Metabolism, 297(2):* E289-E296. doi: 10.1152/ajpendo.00099.2009.

Søreide, K., Weiser, T. G., & Parks, R. W. (2018). Clinical Update on Management of Pancreatic Trauma. *HPB (Oxford), 20(12):* 1099-1108. doi: 10.1016/j.hpb.2018.05.009.

Swami, O. C., & Shah, N. J. (2017). Functional dDyspepsia and the role of digestive enzymes supplement in its therapy. *Int J Basic Clin Pharmacol*, 6(5), 1035-1041. doi: 10.18203/2319-2003.ijbcp20171653.

Unger, R. H., Clark, G. O., Scherer, P. E., & Orci, L. (2010). Lipid Homeostasis, Lipotoxicity and the Metabolic Syndrome. *Biochimica et Biophysica Acta (BBA)-Molecular and Cell Biology of Lipids, 1801(3):* 209-214. doi: 10.1016/j.bbalip.2009.10.006.

Waldmann, E., & Parhofer, K. G. (2019). Apheresis for Severe Hypercholesterolaemia and Elevated Lipoprotein(a). *Pathology, 51(2):* 227-232. doi: 10.1016/j.pathol.2018.10.016.

Watt, M. J., & Steinberg, G. R. (2008). Regulation and Function of Triacylglycerol Lipases in Cellular Metabolism. *Biochemical Journal, 414(3):* 313-325. doi: 10.1042/BJ20080305.

Whitcomb, D. C., Lehman, G. A., Vasileva, G., Malecka-Panas, E., Gubergrits, N., Shen, Y., Sander-Struckmeier, S. and Caras, S., 2010. Pancrelipase Delayed-Release Capsules (CREON) for Exocrine Pancreatic Insufficiency Due to Chronic Pancreatitis or Pancreatic Surgery: A Double-blind Randomized Trial. *Journal of the American College of Gastroenterology, 105(10):* 2276-2286. doi: 10.1038/ajg.2010.201.

Wymann, M. P., & Schneiter, R. (2008). Lipid Signalling in Disease. *Nature Reviews Molecular cell biology, 9(2):* 162-176. doi: https://doi.org/10.1038/nrm2335.

Zhong, L., Feng, Y., Wang, G., Wang, Z., Bilal, M., Lv, H., Jia, S., & Cui, J. (2020). Production and Use of Immobilized Lipases in/on Nanomaterials: A Review from the Waste to Biodiesel Production. *International Journal of Biological Macromolecules, 152:* 207-222. doi: 10.1016/j.ijbiomac.2020.02.258.

Zhu, G., Fang, Q., Zhu, F., Huang, D., & Yang, C. (2021). Structure and Function of Pancreatic Lipase-Related Protein 2 and its Relationship with Pathological States. *Frontiers in Genetics, 12:* 693538. doi: 10.3389/fgene.2021.693538.

Chapter 6

Lipases in Diseased Conditions – Impact on Humans, Animals and Birds

**Sourabh Patil
and Tushar Borse**[*]
School of Biotechnology, Vidya Pratishthan's Arts, Commerce and Science College, Baramati, Pune, Maharashtra, India

Abstract

Cellular enzyme lipases are ubiquitous enzymes produced by specific cells in animals, birds, plants and microorganisms. Lipases hydrolyze carboxylic ester bonds converting triglyceride into fatty acids and glycerol. These enzymes play a crucial role in digestion, transport and processing dietary lipids. Lipases have a wide range of substrates, such as neutral lipids, phospholipids, lysophospholipids, sphingolipids, ether lipids, oxidized lipids, and lipid moieties attached to proteins post-translationally. Lipases are expressed and active in a variety of tissues, lipases like Hepatic lipases, hormone-sensitive lipases, lipoprotein lipase, and pancreatic lipase, found in the liver, adipocytes, vascular endothelial surface, and small intestine, respectively. Lipases have a pivotal role in maintaining health and are implicated in various diseases upon malregulation of its function. Genetic deletion or pharmacological inhibition of some lipases has yielded therapeutic advantages or pathological outcomes in diverse biological scenarios. Some pharmaceutical companies have focused on targeting selected lipase for potential therapeutic applications which results in some drugs which are approved or under clinical trials for obesity, inflammation, cardiovascular diseases, However, the inhibition of specific lipase

---

[*] Corresponding Author's Email: thbvsbt@gmail.com.

In: Lipases and their Role in Health and Disease
Editors: Vasudeo Zambare and Mohd. Fadhil Md. Din
ISBN: 979-8-89113-628-1
© 2024 Nova Science Publishers, Inc.

enzymes by environmental agents has been documented to trigger a range of pathologies, including neurodegenerative conditions, psychotropic effects, and hyperlipidemia.

Keywords: lipases, lipoproteins, cardiovascular diseases, cystic fibrosis, obesity

Introduction

Lipases are ubiquitous enzymes found in all living organisms. Lipases are water soluble, belongs to triacylglycerol ester hydrolases family, and they are capable of facilitating the catabolism and anabolism of long chain triglycerides into fatty acids (FA), diacylglycerol, monoacylglycerol, and glycerol. They are also referred to as carboxylesterase and represented by Enzyme Commission number EC3.1.1.3 (Wu et al., 1996; Lotti and Alberghina, 2007). Most lipases have optimal activity and stability within a pH range of 6.0 to 8.0 and at temperatures ranging from 30°C to 40°C. However, it is important to note that the factors affecting enzyme activity may vary depending on the specific source of the lipases (Sharma et al., 2001).

Lipases are part of pancreatic secretions and are involved in fat metabolism, lipid transport and serve unique functions in varied tissues, viz., hepatic lipase in liver, hormone-sensitive lipases in adipocytes, lipoprotein lipase in endothelial cells, and pancreatic lipase in the small intestine (Pirahanchi and Sharma, 2023). In the liver, they are responsible for the degradation of triglycerides as intermediate-density lipoprotein (IDL). In fat tissue, the hormone-sensitive lipase hydrolyses triglycerides stored within adipocytes. Lipoprotein lipase in the vascular endothelial cells degrades triglycerides which are circulated as chylomicrons and very low-density lipoproteins (VLDLs). Within the small intestine pancreatic lipase is involved in oxidation of dietary triglycerides (Pirahanchi and Sharma, 2023; Young and Zechner, 2013).

Among several lipases, the pancreatic lipases have specific functions and structurally share a common α/β-hydrolase fold and a catalytic triad, where the actual hydrolysis of substrate occurs. The pancreatic lipase gene family is well documented, and it consists of seven mammalian subfamilies: pancreatic lipase, pancreatic lipase related proteins 1 and 2, hepatic lipase, lipoprotein lipase, endothelial lipase, and phosphatidylserine phospholipase A1 (Long and Cravatt, 2011). Additional lipases in mammals like carboxyl ester lipase and gastric lipase are also responsible for lipid digestion. Although lipase enzymes have varied substrate specificity, but it shows major overlapping in functionality among the lipase gene family due to its structural similarities. This has helped in advanced studies for varied applications of lipases in diagnostics, clinical assays, enzyme engineering (Lim et al., 2022).

Lipolysis a biochemical process is responsible for catabolism of triglycerides (TGs), whereas lipophagy is a process involving breakdown of lipid by autophagosis. The FFAs produced serves as an energy source, can also be used for the biosynthesis of lipids, and also act as signal molecule in cell signaling. Apart from the fundamental and physiological role, excess production of FFAs may trigger lipotoxicity which will affect the membrane function, endoplasmic reticulum stress, mitochondrial dysfunction, cell death, and inflammation (Liu and Czaja, 2013). The malfunctioning due to lipase is found in all life forms – humans, animals, birds and plants. The clinical relevance leading to disruption of a process or leading to a specific disease may be unique or share a common feature at genetic and biochemical level. The present chapter will focus on representative diseases in relation to lipases common in Humans, birds and animals.

Lipases in Diseased Conditions In Humans

Lipase is synthesized in the pancreas, mouth, and gastric region facilitating the hydrolysis of dietary fats, transported through the blood into the whole body. Variation in the concentration of lipase level decides the diseased conditions viz., lower levels cause obesity, cystic fibrosis and chronic pancreatitis; whereas higher level causes chronic kidney disease, peptic ulcer, gallbladder disease, intestinal problem, diabetes, alcohol use disorders (Table 1). Very high level of lipases is an indication of acute pancreatitis disease, and its absence leads to Wolman disease. Following are the representative diseases in humans in relation to lipases.

   a. *Obesity:* Obesity is a complex multifactorial disease. Adipocyte hypertrophy is a defining characteristic of obesity in adults. Adipose tissue plays a role in the maintenance of energy balance (Porro et al., 2021). Insulin resistance, which is generated by a diet high in fat has been closely linked to obesity, potent source for the development of diabetes and cardiovascular disease. The adipose tissue functions as the primary location for the storage of TG and the release of free fatty acids, which occurs in reaction to fluctuation in energy requirements (Shoelson et al., 2007).

Table 1. Summary of lipase dependent diseases in humans - symptoms, biochemical level and its treatment

Disease	Lipase level	Symptoms	Treatment	References
Pancreatitis	Elevated	Abdominal pain, nausea, vomiting, fever	Fasting, pain management, enzyme replacement	(Garber et al., 2018; Mederos et al., 2021)
Cystic Fibrosis	Elevated	Respiratory issues, digestive problems	Symptomatic treatment, enzyme supplements	(Konstan et al., 2007; Cystic Fibrosis Foundation et al., 2009)
Pancreatic Cancer	Elevated	Abdominal pain, weight loss, jaundice	Surgery, chemotherapy, radiation	(Lambert et al., 2019; Rahib et al., 2014)
Gallstones	May increase	Abdominal pain, nausea, vomiting	Surgery (cholecystectomy), medication	(Portincasa et al., 2006; Tazuma, 2006)
Hyperlipidaemia	May increase	Often asymptomatic	Lifestyle changes, medication	(Berglund et al., 2012; Reiner, 2017)
Chronic Pancreatitis	Decreased	Abdominal pain, weight loss, fatty stool	Enzyme replacement therapy, pain management	(Fieker et al., 2011)
Pancreatic Insufficiency	Decreased	Malabsorption, nutrient deficiencies	Pancreatic enzyme replacement, dietary adjustments	(Lindkvist, 2013)

Lipoprotein lipase (LPL), a key enzyme for in lipid metabolism and processing, is responsible for the hydrolysis of TG core of circulating TG-rich lipoproteins, chylomicrons, and VLDL (Wang and Eckel, 2009). The products are absorbed and utilized as a source of metabolic energy in peripheral tissues or undergo a process of reconversion into triglycerides. LPL is a rate limiting enzyme, which is found to be regulated at transcriptional, posttranscriptional, and posttranslational levels in a tissue-specific manner. It has also been found to be regulated by the nutritional status and hormonal levels of an individual. There are varieties of associated proteins that interact with LPL responsible to regulate tissue-specific activity of LPL.

The LPL gene resides on the chromosomal location at 8p22 in the human genome. Represent the TG lipase gene family of proteins comprising the hepatic lipase (HL), pancreatic lipase (PL), endothelial lipase (EL). Experimental analysis of the complementary DNA sequence of LPL gene from humans reveals that it consists of 448 amino acids; 10 exons, spanning a distance of approximately 30kb (Brown, 2005).

Variation in expression levels of LPL in organs and tissue disrupts the distribution of plasma TG. Overexpression accumulates TG, develops insulin resistance, favours excessive weight gain, and increases the metabolic rate (Kersten, 2014). Skeletal muscle cells with LPL deletion have reduced TG accumulation and increased insulin action on glucose transport. Variations at molecular and biochemical levels leads to increased lipid partitioning to other tissues, insulin resistance, and obesity. Change in the promoter sequence of the LPL gene has been observed to affect the lipid metabolism responsible for obesity and type 2 diabetes (Wang and Eckel, 2009). These variations are also observed in the tissues and organs causing changes in tissue-specific regulation of LPL responsible for nutrient partitioning when energy intake exceeds energy expenditure.

Insulin and meals raise LPL in adipose tissue, whereas fasting causes a decrease in LPL. In-obesity condition the LPL level per cell is elevated, although it is not affected by meals or insulin. In skeletal muscle, LPL activity is drastically reduced in obesity with weight reduction. LPL plays a crucial role in the metabolism of TG-rich lipoprotein in the heart, a process essential to cardiac function. The exact role of LPL in the brain and spinal cord is not yet unknown but known to show effects as seen in adipose tissues (Klop et al., 2013).

Limited experimental studies are observed in mouse models related to altered LPL expression, facilitating the mechanism of metabolic disorders in humans. However, studies have established that the genetics of LPL deficiency in humans is autosomal recessive due to missense mutations (heterozygous carriers have normal or minimal hypertriglyceridemia) (Chang, 2019).

b. *Atherosclerosis:* The disease is a chronic inflammatory disorder that is triggered and related to accumulation of lipids. It has adverse effects leading to impairment of endothelium, monocytes/ macrophages, T cells, smooth muscle cells, and a regulatory network consisting of growth factors and cytokines. It also affects the pathological processes like vascular wall thickening, hardness, and diminished flexibility (Manduteanu and Simionescu, 2012). At the cellular level, the pathogenesis of atherosclerosis includes the infiltration of many cell types, such as monocytes that undergo differentiation into macrophages within the vascular wall (Fenyo and Gafencu, 2013).

Different concentrations of LPL exert distinct functions and have diverse effects on human health and disease. The effect can be a pro-atherogenic or an anti-atherogenic effect depending on its locations. Mutations leading to impaired lipolysis may be proatherogenic which promotes an alteration in lipoprotein profile followed by uptake and buildup of lipids, allowing macrophages to undergo a transformation into foam cells. Not only alteration in plasma lipoproteins but LPL also affects the obstruction of LDL and VLDL in the arterial intima, thereby increasing its in interactions with the extracellular matrix (Linton et al., 2000). In addition, localised secretion of LPL by macrophages increases the chances of uptake of atherogenic lipoproteins, resulting in elevated formation of foam cells. These foam cells then produce growth factors and metalloproteinases, resulting in the degradation of the extracellular matrix and the formation of lesions in the intima (Lorey et al., 2022).

Lipid accumulation in the arterial wall is a primary factor in atherosclerosis. Most vascular damage is caused by LDL oxidation, which generates proinflammatory substances. Other lipids have similar effects; high plasma VLDL levels increase atherosclerosis risk (Malekmohammad et al., 2021). Studies conducted in 70s and 80s, revealed that an increase in plasma lipoprotein A [Lp(a)] leads to atherosclerotic diseases including coronary heart disease (CHD). Lp(a) has been recognized for its potential proatherogenic features due to its resemblance to low-density lipoprotein (LDL). Additionally, the similarity between apolipoprotein(a) and plasminogen suggests that Lp(a) may potentially exhibit prothrombotic properties. The link between raised Lp(a) and increased risk for CHD events is well-established. The activation of monocytes, important atherogenesis cells, may increase risk. High plasma Lp(a) concentrations affected human blood lymphocyte gene expression. Buechler and colleagues found that pure Lp(a) elevated proinflammatory interleukin-6 (Scanu et al., 1991).

LPL derived from the parenchymal cells of the adipose and muscle tissues acts physiologically at the luminal membrane of blood vessels, where highly charged heparan sulfate proteoglycans attach the enzyme. Studies have focused on monocyte-derived macrophages and have revealed significant insights into the enzyme's pathophysiological actions, which contribute to the formation of foam cells and, consequently, the development of atherosclerosis (Gordts and Esko, 2018). In stark contrast to this, the production of LPL by adipose tissue and muscle exhibits a preventive influence in relation to the development of atherosclerosis. In vivo studies depict that atherosclerosis may develop in subjects with a history of chylomicronemia (impaired fat metabolism) as a result of lipoprotein lipase deficiency (Figure 1) signifying the importance of LPL in protection against atherosclerosis in humans. Based on these discoveries, LPL has been identified as a promising therapeutic target and remains the focus of intense scientific study (Fisher et al., 1997; Y. Li et al., 2014).

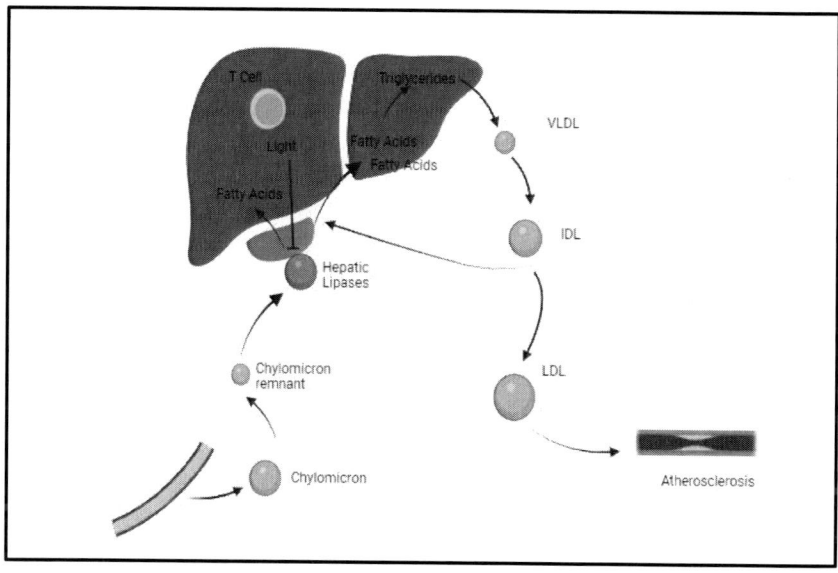

Figure 1. Schematic representation of lipid metabolism by lipoprotein lipases in liver leading to atherosclerosis (VDL- very low-density lipoproteins, IDL- intermediate-density lipoprotein, LDL- low-density lipoproteins).

c. *Wolman's Disease:* Wolman disease is the most severe manifestation of lysosomal acid lipase (LAL) deficiency. It arises as a consequence of genetic abnormalities occurring in the lysosomal acid lipase (LIPA) gene. The LIPA gene encodes the necessary information for the synthesis of the lysosomal acid lipase enzyme. The presence of this enzyme is crucial for the metabolic breakdown of certain lipids in the human body, notably cholesteryl esters, and to lesser extent triglycerides. Without correct values of this enzyme, these fats abnormally accumulate in and harm various tissues and organs of the body (Tylki-Szymańska and Jurecka, 2014).

The occurrence of mutations in the LIPA gene associated with Wolman disease leads to the absence of LIPA enzyme production or the creation of a malfunctioning and non-functional variant of the LIPA enzyme. A novel homozygous mutation in LIPA gene exon 4: converts the glycine amino acid to aspartic acid led to the diagnosis of WD (Ashari et al., 2022). The clinical signs and symptoms of Wolman disease tend to appear in the immediate postnatal period, commonly within the initial weeks following birth. Infants who are affected may exhibit symptoms such as abdominal distention, characterized by bloating or swelling of the stomach, as well as hepatosplenomegaly, which refers to a notable enlargement of the liver and spleen (Aguisanda et al., 2017). Hepatic fibrosis, characterized by the formation of scar tissue in the liver, may also manifest as a consequence. Ascites, the accumulation of fluid in the abdominal cavity, can occur in certain instances.

A potential diagnosis of Wolman disease can be inferred in neonates by observing the presence of distinctive symptoms, including hepatomegaly and gastrointestinal complications. The confirmation of a diagnosis typically involves a comprehensive clinical assessment, an in-depth patient history (including family history), and the utilization of specialized tests that identify the absence or inadequate functioning of the enzyme LIPA in specific cells and tissues inside the body. Additionally, there is the option to conduct molecular genetic testing to identify mutations in the LIPA gene (Hoffman et al., 1993).

 d. *Cholesteryl Ester Storage Diseases (CESD):* Cholesteryl Ester Storage Disease is attributed to a partial insufficiency of the activity of lysosomal acid lipase. The enzyme plays a crucial role in regulation of cholesterol levels within cells, as it is essential for the intracellular breakdown of cholesteryl esters and triglycerides that have been taken up by cells through the process of lipoprotein endocytosis. The accumulation of non-hydrolysed cholesteryl esters and triglycerides in numerous organs, as well as the stimulation of endogenous cholesterol synthesis and LDL production, occurs in cases of LAL activity shortage (Bernstein et al., 2013).

The clinical symptoms of CESD might exhibit considerable heterogeneity. The onset of this condition typically occurs within the initial ten years of life, characterized by disruptions in lipid metabolism, concurrent hepatomegaly, and increased serum concentrations of hepatic transaminases.

During the initial months of life, it is possible to see hepatomegaly, raised cholesterol levels, and increased transaminase activity. Hepatic damage is a gradual condition that ultimately culminates in the development of hepatic fibrosis in all individuals. The manifestation of chronic liver failure and fibrosis can potentially arise during the later stages of childhood or the initial stages of adolescence. Splenomegaly is present in around one-third of patients(Lin et al., 2015). Additional symptoms documented in the existing body of literature include jaundice, recurring stomach pain, gastrointestinal bleeding, and delayed onset of puberty. In the context of CESD, the adrenal glands do not exhibit any alterations in their structure, namely calcification, or their physiological functioning, in contrast to the manifestations shown in individuals with Wolman disease (Reynolds, 2013).

> e. *Cystic Fibrosis:* Cystic fibrosis (CF) is a hereditary disorder that follows an autosomal recessive pattern of inheritance, resulting from genetic abnormalities affecting the cystic fibrosis transmembrane regulator (CFTR) gene. The result is a lack or absence of functional CFTR proteins on the apical membrane of secretory and absorptive epithelial cells in several organs within the gastrointestinal system (Kerem and Kerem, 1996). The impairment of CFTR proteins hinders the trans-epithelial transportation of Cl^- ions via CFTR-associated Cl^- channels, which typically facilitates the excretion of fluid and other ions. The process of dehydration happens in different secretions, such as mucus, at the places that are impacted. This leads to the formation of precipitates and blockage inside the ducts, causing inflammation, fibrosis, and ultimately harm to the organs, especially when digestive enzymes are present (Lukasiak and Zajac, 2021). The pancreas plays a crucial role as the primary organ responsible for the digestion of carbohydrates, proteins, and lipids. This is achieved through the secretion of a diverse array of digestive enzymes into the duodenum. The enzymes mostly encompass pancreatic amylase, protease, lipase, and colipase. Pancreatic acinar cells are responsible for the secretion of pancreatic digesting enzymes in an inactive form into the acinar lumen, which subsequently extends into the pancreatic ducts. The ducts are ductal cells that generate bicarbonate (HCO_3^-) through cAMP stimulation in order to increase alkalinity and dilute the acinar secretions. This process also serves to counteract the acidity of stomach acid within the duodenal lumen.

Salivary amylase, gastric lipase, and pepsin have been identified as potential compensatory mechanisms for the maldigestion and malabsorption observed in CF resulting from pancreatic insufficiency (PI). Research has indicated that gastric lipase is the primary contributor to pre-duodenal lipase activity in humans, with minimal contribution from lingual lipase. These enzymes are secreted by chief cells in the stomach, which also express the CFTR (Li and Somerset, 2014).

Furthermore, it has been observed that gastric lipase, which exhibits an ideal pH of 5.4, demonstrates enhanced enzymatic activity inside the acidic environment of the CF duodenum. It is worth mentioning that gastric lipase is responsible for approximately 90% of the overall lipase activity in the upper small intestine following a meal in individuals with CF and PI.

f. *Acute Pancreatitis*: Acute Pancreatitis is characterized by localized damage due to systemic inflammatory response. The primary triggering occurrence involves the untimely initiation of the trypsinogen enzyme's conversion into trypsin, transpiring within the acinar cell rather than the duct lumen (Bhatia et al., 2005). The phospholipase A2 (PLA2) plays a crucial role in the inflammatory process. It hydrolyzes phospholipids in cell membranes, releasing free fatty acids and lysophospholipids. This leads to cellular injury, inflammation, and the activation of inflammatory mediators, contributing to the severity of pancreatitis. Additionally, the released fatty acids can form toxic byproducts, further aggravating tissue damage (Zhang et al., 2014). The exact mechanism involves a complex interplay of inflammatory pathways and is a subject of ongoing research in understanding the pathophysiology of acute pancreatitis.

In acute pancreatitis, amylase and lipase show significantly increased serum activity. Both are released through urination, but lipases get reabsorbed and its prolonged elevation opens the window for its uses in diagnostics, especially advantageous for delayed presentations (>24 hours) from the onset of pancreatitis. Lipase level starts increasing within 4-8 hours and reaches at peak at 24 hours and also normalized within 7-14 days. Increased lipase (2-50 times the upper limit) demonstrates high clinical sensitivity (80-100%) and specificity (80-100%) for diagnosing acute pancreatitis, depending on the selected cutoff (37).

g. *Cholecystitis:* Cholecystitis is characterized by the inflammatory response of the gallbladder which involves the obstruction of the cystic duct, leading to pathophysiologic changes. This condition may be correlated with or without the existence of gallstones and can also be categorized as either acute or chronic (Knab et al., 2014). The prevalence of cholelithiasis in individuals diagnosed with acute cholecystitis is estimated to be greater than or equal to 95% (Gallaher and Charles, 2022). Acute inflammation occurs when a stone becomes impacted in the cystic duct, causing chronic obstruction. The stagnation of bile induces the liberation of inflammatory enzymes, such as phospholipase A. This enzyme facilitates the conversion of lecithin into lysolecithin, which in turn has the potential to mediate the inflammatory response (Sjödahl and Tagesson, 1983). Mild cholestatic abnormalities, characterized by bilirubin levels of up to 4 mg/dL and slightly increased alkaline phosphatase, are frequently observed (Nguyen et al., 2014). Greater and more noticeable elevations, particularly when lipase is increased by more than three times, indicate the presence of bile duct obstruction. The presence of a stone in the biliary tract leads to an elevation in aminotransferases, specifically alanine aminotransferase and aspartate aminotransferase (Besinger and Stehman, 2016).

However, lipases are not directly implicated in the development of this condition with reference to cholecystitis. But it may indirectly be affected by cholecystitis, and their levels can be influenced by the digestive changes associated with gallbladder inflammation (Rodríguez-Antonio et al., 2020). Cholecystitis can impact the normal storage and release of bile (bile salts and lipases) from the gallbladder. Any disruption in the normal flow of bile can affect the digestion and breakdown of lipids. Cholecystitis can lead to impaired bile flow and affect the release of pancreatic lipase enzymes, potentially influencing lipid digestion. Clinical testing can involve blood tests to assess liver function, including levels of liver enzymes (Wang, 2008). While lipase levels are not typically used to diagnose cholecystitis, but elevated levels affect the pancreas (pancreatitis), which can sometimes occur concurrently with gallbladder issues.

Lipases in Diseased Conditions in Animals

Similar to humans, lipases play a crucial role in the breakdown and digestion of fats (lipids) in animals. Lipase-related diseases affect varied animal species, impacting lipid metabolism and digestion. As compared to human lipase-related diseases are not as extensively studied in animals, but there are conditions similar to that of humans that can affect lipase function in different species of animals. Few notable diseases are as discussed below.

a. *Pancreatitis:* Pancreatitis is a prevalent inflammatory disorder observed in both dogs and cats. The condition can be either acute or chronic and may present as subclinical or exhibit a range of clinical indications. The diagnosis of pancreatitis involves the integration of clinical presentation, imaging results, and serum lipase immunoreactivity levels (Watson, 2015). The etiology of pancreatitis in canines and felines is mostly unknown, with the majority of cases being classified as idiopathic (Xenoulis, 2015). Dietary indiscretion is widely regarded as a prevalent risk factor in dogs. At severe level serum concentrations of 500 mg/dL or above a condition called hypertriglyceridemia is a risk factor for pancreatitis in dogs, but not in cats. Several studies have identified hyperadrenocorticism as a potential risk factor for pancreatitis in dogs (Wang et al., 2022). Pancreatitis can also be induced by severe blunt trauma, which may occur as a result of a road accident or in felines experiencing high-rise syndrome.

Pancreatitis has been documented in dogs afflicted with either Babesia canis or Leishmania infection. The most significant pathogens in cats are *Toxoplasma gondii*, *Amphimerus pseudofelineus*, and feline infectious peritonitis (Zoran, 2006). Research studies have proposed 1,2-o-dilauryl-rac-glycero-3-glutaric acid-(6'-methylresorufin) ester (DGGR) and triolein as more effective substrates for assessing pancreatic lipase activity in serum (Graca et al., 2005). Assessment of pancreatic lipase immunoreactivity (PLI) is exclusively intended for quantifying the concentration of pancreatic lipase in serum, making it the most precise diagnostic method for pancreatitis exhibiting high levels of sensitivity.

b. *Inflammatory Bowel Disease:* Inflammatory bowel disease (IBD) is a complex ailment affecting dogs and cats, marked by persistent enteropathies that have a substantial impact on the overall well-being of the animals. Typically, these enteropathies are classified as diet responsive, antibiotic responsive, steroid responsive, or refractory, irrespective of the use of immunosuppressive medications (referred to as idiopathic IBD) (Yogeshpriya et al., 2017). Histologically, it is possible for either the small intestine, large intestine, or both to be impacted. Lymphocytes and plasma cells are the predominant cellular infiltrates observed in the lamina propria of the gastrointestinal (GI) tract, whereas eosinophils, macrophages, and neutrophils are present but with lower frequency.

Cats diagnosed with IBD may exhibit elevated levels of serum pancreatic lipase, which may indicate the presence of pancreatitis. According to a recent analysis, it seems that this relationship does not exert any discernible influence on clinical outcome. Nevertheless, there is a correlation between elevated levels of pancreatic lipase in the serum of dogs with IBD and a less favorable clinical prognosis (Xenoulis, 2015).

c. *Malabsorption Syndrome:* The significance of lipase in malabsorption syndromes in cats and dogs cannot be overruled due to its importance in the process of breaking down and assimilating dietary fats. This syndrome in animals can be attributed to a range of factors, with lipase deficits or dysfunctions playing a substantial role in the compromised absorption of fats (Williams, 2005). Exocrine Pancreatic Insufficiency (EPI) is a prevalent etiology of malabsorption in canines and, to a lesser extent, in felines. EPI is characterized by a scarcity of pancreatic enzymes, specifically lipase, resulting in compromised fat digestion. Insufficient lipase activity can give rise to inadequate fat digestion, hence causing the presence of undigested lipids in the stools, a condition known as steatorrhea (Xenoulis, 2020). Malabsorption syndromes in canines and felines can manifest through several clinical indications, including but not limited to weight loss, diarrhea, steatorrhea, and suboptimal coat condition (Westermarck and Wiberg, 2012).

d. *Cystic fibrosis:* CF, a condition developed in humans, is rarely found in animals. However, studies with animal models, such as mice, ferrets, and pigs, CF is found to affect the pancreas, liver, and gallbladder, with characteristic acidic, dehydrated, and protein-rich secretions.

Mutations in CFTR genes in pigs are localized on pancreatic ducts. The histopathological characteristics of the liver and pancreas closely resemble those observed in humans with the same condition (Uc et al., 2012). The CF model pig experience pancreatic insufficiency due to thickened mucus blocking the pancreatic ducts is characterized by reduced release of digestive enzymes, including lipase. The absence or reduction of lipase and other pancreatic enzymes can lead to difficulties in digesting dietary fats contributing to malabsorption of fats, leading to nutritional deficiencies and related health issues (Meyerholz et al., 2010). The porcine model, with its similarities to humans in terms of pancreatic and hepatic histopathology, offers a valuable opportunity to investigate the specific effects of CF mutations on lipase function and lipid metabolism.

The new animal model developed Donryu rats carry genetic lipid storage disease, resembling the human Wolman's disease. It is an autosomal recessive disease and characterized by hepatosplenomegaly, lymph node enlargement, and thickened, dilated intestine, which ultimately results in the formation of morphologically characteristic foam cells in livers and spleens. Biochemical studies on spleen and liver tissues showed massive accumulation of esterified cholesterol and triglycerides, and deficiency of acid lipase for [^{14}C]-cholesteryl oleate(Yoshida and Kuriyama, 1990). This model has contributed greatly to the clarification of the physiological and pathological roles of lysosomal acid lipase in the metabolism of lipoproteins and cholesterol, and of the pathogenesis of atherosclerosis.

Lipases in Diseased Conditions in Birds

As studied in animals and humans, lipase also plays an essential role for the proper digestion and absorption of dietary fats. Though there are no specific lipase-related diseases cited in birds, issues related to lipid metabolism or digestion can arise and impact avian health. Let's try to understand a few conditions related to lipid metabolism in birds.

Fatty Liver Disease

Fatty liver disease also referred to as fatty liver syndrome, and hepatic lipidosis, a condition observed in birds signifies a gradual accumulation of fat, specifically large vacuoles of triglyceride fat, within the normal liver cells (Visscher et al., 2017). These abnormal cells lose their ability to efficiently carry out the liver's functions, and as liver cells perish, they are replaced by scar tissue or fibrous connective tissue. As this process unfolds over time, the liver's functionality diminishes, leading to the manifestation of signs indicative of liver disease in the bird. For example the Amazon Parrots, Turkey, Quakers, Budgerigars, Cockatiels, Lovebirds, and Rose-breasted cockatoos are the species most commonly affected by hepatic lipidosis (Miesle, 2022). Female birds are more prone to this condition, possibly linked to hormonal activity during reproduction. Birds with a high-seed diet are more susceptible to obesity, as seeds are rich in fat but deficient in essential nutrients like biotin, choline, and methionine.

Long-term exposure to toxins, particularly aflatoxins found in certain peanuts and corn products, can lead to repeated liver damage and the onset of hepatic lipidosis. The toxins may accumulate in the bloodstream, leading to central nervous system signs like disorientation or seizures. The progression of liver disease is typically gradual, and the bird may appear suddenly ill when the liver tissue is replaced by fat over time (Hamilton and Garlich, 1971). Enlarged liver restricts breathing by compromising the body cavity space, leading to a distended abdomen (liver may be visible beneath the skin below the keel). Birds with hepatic lipidosis may exhibit diarrhea, and their droppings may take on a greenish hue due to the excretion of biliverdin. Some display poor feather quality, bleeding tendencies and clotting issues.

Malabsorption Syndrome

The common symptoms linked to the disease are the abnormalities in the absorption or digestion of nutrients in the intestines, leading to the term "malabsorption syndrome." There is disruption in the production or activity of lipase, leading to difficulties in fat digestion and absorption. Birds affected with the disease experience diminished levels of carotenoids in both the liver and serum. Additionally, there is a reduction in serum levels of vitamins A, E, and D. The pale coloration observed in the legs and beak is attributed to the decreased absorption of carotenoids (Ruff, 1982). Malabsorption of various nutrients, as well as decreased utilization of fats and proteins, is significantly evident in affected birds when compared to healthy controls. Impaired lipid absorption is also noted, along with noticeable fluctuations in the utilization of different dietary fatty acids. Lesions are also a main sign of malabsorption syndrome; in poultry it involves enlarged proventriculi, reduced gizzard size, pancreatic, thymic, and bursal atrophy, along with the presence of orange mucus in the small-intestinal lumen (Page et al., 1982).

Conclusion

Lipases are produced and found in liver, adrenal glands, adipose tissues and ovaries, but more than 95% of the total activity is contributed by the hepatic lipases for plasma lipoprotein metabolism. They have also been found to be derived from macrophages and have been found important in atherosclerotic individuals. Biochemical significance of lipase as a hydrolases cleaving the carboxylic ester bonds converting TG into chylomicrons and VLDL is well documented. Lipases help in maintaining normal levels of fat and its forms in a tissue specific manner with reference to its energy requirements and functioning of the tissue, organ or the cell. Owing to these properties it has gained clinical significance and is regarded as a promising candidate for designing diagnostic methods, drug designing or therapy to treat the lipase dependent diseases. Elaborative and extensive studies in humans, animals and birds have helped researches to gain insights to explore lipases and its role in specific diseased condition. Deficiency of the lipases affects the process of lipolysis, transport of FFAs and may lead to leads to malabsorption of fats and fat-soluble vitamins. Serum lipid levels in varied diseases may vary, but most children and adults present elevated and reduced LDL/HDL levels. Organ abnormalities in these diseases are well studied, the rate of progression and presentation of symptoms are not always consistent. Lipases have been found to be a good candidate for enzyme engineering and a target for drug development for diseases like cystic fibrosis, atherosclerosis, Hyperlipidaemia, Chronic Pancreatitis, Wolman's disease etc. Comparative studies of genetic variants in animal and bird models also support to the significance of lipases in lipoprotein metabolism and diseases prevalent in humans. Although the data available in animals and birds is limited but comparative studies in humans suggest a similarity in role of lipases in fat metabolism as well as transport and diseased conditions.

References

Aguisanda, F., Thorne, N., & Zheng, W. (2017). Targeting Wolman Disease and Cholesteryl Ester Storage Disease: Disease Pathogenesis and Therapeutic Development. *Current Chemical Genomics and Translational Medicine, 11:* 1–18. doi: 10.2174/2213988501711010001.

Ashari, K. A., Azari-Yam, A., Shahrooei, M., & Ziaee, V. (2022). Wolman disease presenting with hemophagocytic lymphohistiocytosis syndrome and a novel LIPA gene variant: a case report and review of the literature. *Journal of Medical Case Reports, 17(1):* 369. doi: 10.1186/s13256-023-04116-4.

Berglund, L., Brunzell, J. D., Goldberg, A. C., Goldberg, I. J., Sacks, F., Murad, M. H., Stalenhoef, A. F., & Endocrine Society. (2012). Evaluation and Treatment of Hypertriglyceridemia: An Endocrine Society Clinical Practice Guideline. *The Journal of Clinical Endocrinology and Metabolism, 97(9):* 2969–2989. doi: 10.1210/jc.2011-3213.

Bernstein, D. L., Hülkova, H., Bialer, M. G., & Desnick, R. J. (2013). Cholesteryl Ester Storage Disease: Review of the Findings in 135 Reported Patients with an Underdiagnosed Disease. *Journal of Hepatology, 58(6):* 1230–1243. doi: 10.1016/j.jhep.2013.02.014.

Besinger, B., & Stehman, C. R. (2016). Pancreatitis and Cholecystitis. *Alcohol, 25(35),* 9–12.

Bhatia, M., Wong, F. L., Cao, Y., Lau, H. Y., Huang, J., Puneet, P., & Chevali, L. (2005). Pathophysiology of Acute Pancreatitis. *Pancreatology, 5(2–3):* 132–144. doi: 10.1159/000085265.

Brown, R. J. (2005). *Functional Analysis of the Carboxyl Terminus of Human Hepatic Lipase* (Doctoral Thesis), University of Ottawa (Canada). http://hdl.handle.net/10393/29201.. doi: 10.20381/ruor-12818.

Chang, C. L. (2019). Lipoprotein Lipase: New Roles for an "Old" Enzyme. *Current Opinion in Clinical Nutrition and Metabolic Care, 22(2):* 111–115. doi: 10.1097/MCO.0000000000000536.

Cystic Fibrosis Foundation, Borowitz, D., Parad, R. B., Sharp, J. K., Sabadosa, K. A., Robinson, K. A., Rock, M. J., Farrell, P. M., Sontag, M. K., Rosenfeld, M., Davis, S. D., Marshall, B. C., & Accurso, F. J. (2009). Cystic Fibrosis Foundation Practice Guidelines for the Management of Infants with Cystic Fibrosis Transmembrane Conductance Regulator-Related Metabolic Syndrome During the First Two Years of life and beyond. *The Journal of Pediatrics, 155(6 Suppl):* S106-16. doi: 10.1016/j.jpeds.2009.09.003.

Fenyo, I. M., & Gafencu, A. V. (2013). The Involvement of the Monocytes/Macrophages in Chronic Inflammation Associated with Atherosclerosis. *Immunobiology, 218(11):* 1376–1384. doi: 10.1016/j.imbio.2013.06.005.

Fieker, A., Philpott, J., & Armand, M. (2011). Enzyme Replacement Therapy for Pancreatic Insufficiency: Present and Future. *Clinical and Experimental Gastroenterology, 4:* 55–73. doi: 10.2147/ceg.s17634.

Fisher, R. M., Humphries, S. E., & Talmud, P. J. (1997). Common Variation in the Lipoprotein Lipase Gene: Effects on Plasma Lipids and Risk of Atherosclerosis. *Atherosclerosis, 135(2):* 145–159. doi: 10.1016/S0021-9150(97)00199-8.

Gallaher, J. R., & Charles, A. (2022). Acute Cholecystitis: A Review. *The Journal of the American Medical Association, 327(10):* 965–975. doi: 10.1001/jama.2022.2350.

Garber, A., Frakes, C., Arora, Z., & Chahal, P. (2018). Mechanisms and Management of Acute Pancreatitis. *Gastroenterology Research and Practice, 2018:* 6218798. doi: 10.1155/2018/6218798.

Gordts, P. L. S. M., & Esko, J. D. (2018). The Heparan Sulfate Proteoglycan Grip on Hyperlipidemia and Atherosclerosis. *Matrix Biology, 71–72:* 262–282. doi: 10.1016/j.matbio.2018.05.010.

Graca, R., Messick, J., McCullough, S., Barger, A., & Hoffmann, W. (2005). Validation and Diagnostic Efficacy of a Lipase Assay Using the Substrate 1,2-o-Dilauryl-rac-Glycero Glutaric acid-(6' Methyl Resorufin)-Ester for the Diagnosis of Acute Pancreatitis in Dogs. *Veterinary Clinical Pathology / American Society for Veterinary Clinical Pathology, 34(1):* 39–43. doi: 10.1111/j.1939-165X.2005.tb00007.x .

Hamilton, P. B., & Garlich, J. D. (1971). Aflatoxin as a Possible Cause of Fatty Liver Syndrome in Laying Hens. *Poultry Science, 50(3):* 800–804. doi: 10.3382/ps.0500800.

Hoffman, E. P., Barr, M. L., Giovanni, M. A., & Murray, M. F. (1993). Lysosomal Acid Lipase Deficiency. In Adam, M. P., Feldman, J., Mirzaa G. M., Pagon, R. A., Wallace, S. E., Bean L. J. H. Gripp, K. W., & Amemiya, A. (Eds.), *GeneReviews*, Seattle (WA): University of Washington, Seattle.

Kerem, B., & Kerem, E. (1996). The Molecular Basis for Disease Variability in Cystic Fibrosis. *European Journal of Human Genetics, 4(2):* 65–73. doi: 10.1159/000472174.

Kersten, S. (2014). Physiological Regulation of Lipoprotein Lipase. *Biochimica et Biophysica Acta, 1841(7):* 919–933. doi: 10.1016/j.bbalip.2014.03.013.

Klop, B., Elte, J. W. F., & Cabezas, M. C. (2013). Dyslipidemia in Obesity: Mechanisms and Potential Targets. *Nutrients, 5(4):* 1218–1240. doi: 10.3390/nu5041218.

Knab, L. M., Boller, A.-M., & Mahvi, D. M. (2014). Cholecystitis. *The Surgical Clinics of North America, 94(2):* 455–470. doi: 10.1016/j.suc.2014.01.005.

Konstan, M. W., Schluchter, M. D., Xue, W., & Davis, P. B. (2007). Clinical Use of Ibuprofen is Associated with Slower FEV1 Decline in Children with Cystic Fibrosis. *American Journal of Respiratory and Critical Care Medicine, 176(11):* 1084–1089. doi: 10.1164/rccm.200702-181OC.

Lambert, A., Schwarz, L., Borbath, I., Henry, A., Van Laethem, J. L., Malka, D., Ducreux, M., & Conroy, T. (2019). An Update on Treatment Options for Pancreatic Adenocarcinoma. *Therapeutic Advances in Medical Oncology, 11:* 1758835919875568. doi: 10.1177/1758835919875568.

Lim, S. Y., Steiner, J. M., & Cridge, H. (2022). Lipases: It's Not Just Pancreatic Lipase! *American Journal of Veterinary Research, 83(8).* doi: 10.2460/ajvr.22.03.0048.

Lindkvist, B. (2013). Diagnosis and Treatment of Pancreatic Exocrine Insufficiency. *World Journal of Gastroenterology, 19*(42): 7258–7266. doi: 10.3748/wjg.v19.i42.7258.

Linton, M. F., Yancey, P. G., Davies, S. S., Jerome, W. G. (Jay), Linton, E. F., & Vickers, K. C. (2000). The Role of Lipids and Lipoproteins in Atherosclerosis. In Feingold, K. R., Anawalt, B., Blackman, M. R., Boyce, A., Chrousos, G., Corpas, E., de Herder, W. W., Dhatariya, K., Dungan, K., Hofland, J., Kalra, S., Kaltsas, G., Kapoor, N., Koch, C., Kopp, P., Korbonits, M., Kovacs, C. S., Kuohung, W., Laferrère, B., Levy, M., McGee, E. A., McLachlan, R., New, M., Purnell, J., Sahay, R., Shah, A. S., Singer, F., Sperling, M. A., Stratakis, C. A., Trence, D. L., & Wilson, D. P. (Eds.), *Endotext*. South Dartmouth (MA): MDText.com, Inc.

Lin, P., Raikar, S., Jimenez, J., Conard, K., & Furuya, K. N. (2015). Novel Mutation in a Patient with Cholesterol Ester Storage Disease. *Case Reports in Genetics, 2015*: 347342. doi: 10.1155/2015/347342.

Liu, K., & Czaja, M. J. (2013). Regulation of Lipid Stores and Metabolism by Lipophagy. *Cell Death and Differentiation, 20(1):* 3–11. doi: 10.1038/cdd.2012.63.

Li, L., & Somerset, S. (2014). Digestive System Dysfunction in Cystic Fibrosis: Challenges for Nutrition Therapy. *Digestive and Liver Disease, 46(10):* 865–874. doi: 10.1016/j.dld.2014.06.011.

Li, Y., He, P. P., Zhang, D.W., Zheng, X. L., Cayabyab, F. S., Yin, W. D., & Tang, C. K. (2014). Lipoprotein Lipase: From Gene to Atherosclerosis. *Atherosclerosis, 237(2):* 597–608. doi: 10.1016/j.atherosclerosis.2014.10.016.

Long, J. Z., & Cravatt, B. F. (2011). The Metabolic Serine Hydrolases and Their Functions in Mammalian Physiology and Disease. *Chemical Reviews, 111(10):* 6022–6063. doi: 10.1021/cr200075y.

Lorey, M. B., Öörni, K., & Kovanen, P. T. (2022). Modified Lipoproteins Induce Arterial Wall Inflammation during Atherogenesis. *Frontiers in Cardiovascular Medicine, 9*: 841545. doi: 10.3389/fcvm.2022.841545.

Lotti, M., & Alberghina, L. (2007). Lipases: Molecular Structure and Function. In Polaina, J., & MacCabe, A. P. (Eds.), *Industrial Enzymes* (263–281). Dordrecht: Springer netherlands. doi: 10.1007/1-4020-5377-0_16.

Lukasiak, A., & Zajac, M. (2021). The Distribution and Role of the CFTR Protein in the Intracellular Compartments. *Membranes, 11(11):* 804. doi: 10.3390/membranes 11110804.

Malekmohammad, K., Bezsonov, E. E., & Rafieian-Kopaei, M. (2021). Role of Lipid Accumulation and Inflammation in Atherosclerosis: Focus on Molecular and Cellular Mechanisms. *Frontiers in Cardiovascular Medicine, 8*: 707529. doi: 10.3389/fcvm. 2021.707529.

Manduteanu, I., & Simionescu, M. (2012). Inflammation in Atherosclerosis: A Cause or A Result of Vascular Disorders? *Journal of Cellular and Molecular Medicine, 16*(9): 1978–1990. doi: 10.1111/j.1582-4934.2012.01552.x.

Mederos, M. A., Reber, H. A., & Girgis, M. D. (2021). Acute Pancreatitis: A Review. *The Journal of The American Medical Association, 325(4):* 382–390. doi: 10.1001/jama. 2020.20317.

Meyerholz, D. K., Stoltz, D. A., Pezzulo, A. A., & Welsh, M. J. (2010). Pathology of Gastrointestinal Organs in a Porcine Model of Cystic Fibrosis. *The American Journal of Pathology, 176(3):* 1377–1389. doi: 10.2353/ajpath.2010.090849.

Miesle, J. (2022). *Overview of Avian Geriatric Disorders with Emphasis on Psittacines.* https://www.academia.edu/69299782/Overview_of_Avian_Geriatric_Disorders_with_Emphasis_on_Psittacines.

Nguyen, K. D., Sundaram, V., & Ayoub, W. S. (2014). Atypical Causes of Cholestasis. *World Journal of Gastroenterology, 20(28):* 9418–9426. doi: 10.3748/wjg.v20.i28.9418.

Page, R. K., Fletcher, O. J., Rowland, G. N., Gaudry, D., & Villegas, P. (1982). Malabsorption Syndrome in Broiler Chickens. *Avian Diseases, 26(3):* 618–624.

Pirahanchi, Y., & Sharma, S. (2023). Biochemistry, Lipase. In *StatPearls.* Treasure Island (FL): StatPearls Publishing.

Porro, S., Genchi, V. A., Cignarelli, A., Natalicchio, A., Laviola, L., Giorgino, F., & Perrini, S. (2021). Dysmetabolic Adipose Tissue in Obesity: Morphological and Functional Characteristics of Adipose Stem Cells and Mature Adipocytes in Healthy and Unhealthy Obese Subjects. *Journal of Endocrinological Investigation, 44(5):* 921–941. doi: 10.1007/s40618-020-01446-8.

Portincasa, P., Moschetta, A., & Palasciano, G. (2006). Cholesterol Gallstone Disease. *The Lancet, 368(9531):* 230–239. doi: https://doi.org/10.1016/S0140-6736(06)69044-2.

Rahib, L., Smith, B. D., Aizenberg, R., Rosenzweig, A. B., Fleshman, J. M., & Matrisian, L. M. (2014). Projecting Cancer Incidence and Deaths to 2030: The Unexpected Burden of Thyroid, Liver, and Pancreas Cancers in the United States. *Cancer Research, 74(11):* 2913–2921. doi: 10.1158/0008-5472.CAN-14-0155.

Reiner, Ž. (2017). Hypertriglyceridaemia and Risk of Coronary Artery Disease. *Nature Reviews Cardiology, 14(7):* 401–411. doi: 10.1038/nrcardio.2017.31.

Reynolds, T. (2013). Cholesteryl Ester Storage Disease: A Rare and Possibly Treatable Cause of Premature Vascular Disease and Cirrhosis. *Journal of Clinical Pathology,* 6610.1136/jclinpath-2012-201302.

Rodríguez-Antonio, I., López-Sánchez, G. N., Garrido-Camacho, V. Y., Uribe, M., Chávez-Tapia, N. C., & Nuño-Lámbarri, N. (2020). Cholecystectomy As a Risk Factor for Non-Alcoholic Fatty Liver Disease Development. *HPB: The Official Journal of the International Hepato Pancreato Biliary Association, 22(11):* 1513–1520. doi: 10.1016/j.hpb.2020.07.011.

Ruff, M. D. (1982). Nutrient Absorption and Changes in Blood Plasma of Stunted Broilers. *Avian Diseases, 26(4):* 852–859.

Scanu, A. M., Lawn, R. M., & Berg, K. (1991). Lipoprotein(a) and Atherosclerosis. *Annals of Internal Medicine, 115(3):* 209–218. doi: 10.7326/0003-4819-115-3-209.

Sharma, R., Chisti, Y., & Banerjee, U. C. (2001). Production, Purification, Characterization, and Applications of lipases. *Biotechnology Advances, 19(8):* 627–662. doi: 10.1016/s0734-9750(01)00086-6.

Shoelson, S. E., Herrero, L., & Naaz, A. (2007). Obesity, Inflammation, and Insulin Resistance. *Gastroenterology, 132(6):* 2169–2180. doi: 10.1053/j.gastro.2007.03.059.

Sjödahl, R., & Tagesson, C. (1983). On the Development of Primary Acute Cholecystitis. *Scandinavian Journal of Gastroenterology, 18(5):* 577–579. doi: 10.3109/00365528309181641.

Tazuma, S. (2006). Gallstone Disease: Epidemiology, Pathogenesis, and Classification of Biliary Stones (Common Bile Duct and Intrahepatic). *Best Practice & Research. Clinical Gastroenterology, 20(6):* 1075–1083. doi: 10.1016/j.bpg.2006.05.009.

Tylki-Szymańska, A., & Jurecka, A. (2014). Lysosomal Acid Lipase Deficiency: Wolman Disease and Cholesteryl Ester Storage Disease. *Pril (Makedon Akad Nauk Umet Odd Med Nauki), 35(1):* 99–106.

Uc, A., Giriyappa, R., Meyerholz, D. K., Griffin, M., Ostedgaard, L. S., Tang, X. X., Abu-El-Haija, M., Stoltz, D. A., Ludwig, P., Pezzulo, A., Abu-El-Haija, M., Taft, P., & Welsh, M. J. (2012). Pancreatic and biliary secretion are both altered in cystic fibrosis pigs. *American Journal of Physiology. Gastrointestinal and Liver Physiology, 303(8):* G961-8. doi: 10.1152/ajpgi.00030.2012.

Visscher, C., Middendorf, L., Günther, R., Engels, A., Leibfacher, C., Möhle, H., Düngelhoef, K., Weier, S., Haider, W., & Radko, D. (2017). Fat content, fatty acid pattern and iron content in livers of turkeys with hepatic lipidosis. *Lipids in Health and Disease, 16(1):* 98. doi: 10.1186/s12944-017-0484-8.

Wang, Helen, H. (2008). Molecular Pathophysiology and Physical Chemistry of Cholesterol Gallstones. *Frontiers in Bioscience, 13(13):* 401. doi: /10.2741/2688.

Wang, Hong, & Eckel, R. H. (2009). Lipoprotein Lipase: from Gene to Obesity. *American Journal of Physiology. Endocrinology and Metabolism, 297(2):* E271-88. doi: 10.1152/ajpendo.90920.2008.

Wang, L., Xu, T., Wang, R., Wang, X., & Wu, D. (2022). Hypertriglyceridemia Acute Pancreatitis: Animal Experiment Research. *Digestive Diseases and Sciences, 67(3):* 761–772. doi: https://doi.org/10.1007/s10620-021-06928-0.

Watson, P. (2015). Pancreatitis in Dogs and Cats: Definitions and Pathophysiology. *The Journal of Small Animal Practice, 56(1):* 3–12. doi: 10.1111/jsap.12293.

Westermarck, E., & Wiberg, M. (2012). Exocrine Pancreatic Insufficiency in the Dog: Historical Background, Diagnosis, and Treatment. *Topics in Companion Animal Medicine, 27(3):* 96–103. doi: 10.1053/j.tcam.2012.05.002.

Williams, D. A. (2005). Malabsorption. In Hall, Simpson, & Williams (Eds.), *BSAVA Manual of Canine and Feline Gastroenterology*, 87–90. British Small Animal Veterinary Association. doi: 10.22233/9781910443361.10.

Wu, X. Y., Jääskeläinen, S., & Linko, Y. Y. (1996). An Investigation of Crude Lipases for Hydrolysis, Esterification, and Transesterification. *Enzyme and Microbial Technology, 19(3):* 226–231. doi: 10.1016/0141-0229(95)00239-1.

Xenoulis, P. G. (2020). Exocrine Pancreatic Insufficiency in Dogs and Cats. In Bruyette, D. S., Bexfield, N., Chretin, J. D., Kidd, L., Kube, S., Langston, C., Owen, T. J., Oyama, M. A., Peterson, N., Reiter, L. V., Rozanski, E. A., Ruaux, C., & Torres, S. M. F. (Eds.), *Clinical Small Animal Internal Medicine*, 583–590. Wiley. doi: 10.1002/9781119501237.ch54.

Xenoulis, P G. (2015). Diagnosis of Pancreatitis in Dogs and Cats. *The Journal of Small Animal Practice, 56(1):* 13–26. DOI: 10.1111/jsap.12274.

Yogeshpriya, S., Veeraselvam, M., Krishnakumar, S., Arulkumar, T., Jayalakshmi, K., Saravanan, M., & Selvaraj, P. (2017). Technical Review on Inflammatory Bowel Disease in Dogs and Cats. *International Journal of Science, Environment and Technology, 6(3):* 1833–1842.

Yoshida, H., & Kuriyama, M. (1990). Genetic Lipid Storage Disease with Lysosomal Acid Lipase Deficiency in Rats. *Laboratory Animal Science, 40(5):* 486–489.

Young, S. G., & Zechner, R. (2013). Biochemistry and Pathophysiology of Intravascular and Intracellular Lipolysis. *Genes & Development, 27(5):* 459–484. doi: 10.1101/gad.209296.112.

Zhang, M. S., Zhang, K. J., Zhang, J., Jiao, X. L., Chen, D., & Zhang, D. L. (2014). Phospholipases A-II (PLA2-II) Induces Acute Pancreatitis Through Activation of the Transcription Factor NF-kappaB. *European Review for Medical & Pharmacological Sciences, 18(8):* 1163-1169.

Zoran, D. L. (2006). Pancreatitis in Cats: Diagnosis and Management of a Challenging Disease. *Journal of the American Animal Hospital Association, 42(1):* 1–9. doi: 10.5326/0420001.

About the Editors

Dr. Vasudeo P. Zambare
R&D and Technical Head, Balaji Enzyme and Chemical Ltd, Mumbai, India

Dr. Vasudeo Zambare is a R&D and Technical Head at Balaji Enzyme and Chemical, Pvt. Ltd. He received his bachelor's and master's degrees in chemistry and biochemistry from North Maharashtra University in Jalgaon, India, and his Ph.D. degree in biochemistry from Agharkar Research Institute (affiliated with Pune University) in Pune, India, in 2007. Dr. Zambare's career profile is linked with microbial fermentation and industrial enzymes.

Dr. Zambare started his career as a microbiologist in the Indian food industry and holds the positions of Senior Manager, General Manager, and Vice President for several Indian biotechnology industries. He also holds an academic position as the Dean of the School of Science at a UGC-approved private university in India.

He is a multi-skilled researcher with biorefinery/industry experience in the US, Canada, the EU, Malaysia, and India. He has worked as a Research Scientist at the Center for Bioprocessing Research & Development at the South Dakota School of Mines and Technology in the USA; a Post Doctoral Fellow at the Biorefining Research Institute at Lakehead University in Canada; a Project Manager on EU-Horizon 2020 project at Celignis Limited in Ireland; and a UTM Research Fellow at Universiti Teknologi Malaysia in Malaysia. He has three patents and more than 220 technical and scientific contributions including peer-reviewed research papers, review papers, books, book chapters, proceedings, popular articles, and nucleotide sequences as well as more than ninety peer-reviewed journal articles. Dr. Zambare is a Fellow of six international societies and the Editor-in-Chief of seven international journals. He is an Associate Editor and editorial board member of several international biotechnology journals and a member of multiple advisory boards and professional societies. In 2008, Dr. Zambare was awarded the Best Scientist Award by the Rotary Club at Agharkar Research Institute in India and was appointed as the Indian Chapter Chairperson for the Biotechnology Society of Nepal. He has developed several bioprocesses for the leather, textile, paper and pulp, and biofuel industries, and has more than twenty years of experience in fermentation process development, assays, and analytical methods to solve complex research problems with potential commercial applications in the biofuel, food, and pharmaceutical industries. He also has expertise in industrial enzymes, probiotics, extremophiles, biofertilizers, biopesticides, and waste management.

Professor Ir. Ts. Dr. Mohd. Fadhil Md. Din

Center for Environmental Sustainability and Water Security (IPASA) & Department of Water and Environmental Engineering, School of Civil Engineering, Universiti Teknologi Malaysia, Bahru, Malaysia
Director, Campus Sustainability, Universiti Teknologi Malaysia, Bahru, Malaysia

Professor Mohd Fadhil Md Din graduated from Universiti Teknologi Malaysia (UTM) in 1999 and obtained an MEng in 2001. He further studied under a double program between UTM and Delft University of Technology in the Netherlands. Professor Din has more than 280 technical and scientific publications to his credit. He is a recipient of several funding grants and prestigious awards. Professor Din is a director of campus sustainability, working for an environmental and sustainable operational framework on campus. His expertise is in environmental sciences and application, environmental management systems (EMS), environmental impact assessments (EIA), risk management, and transportation projects (mobility transformational agenda). Professor Din is involved in undergraduate and postgraduate courses in environmental science and engineering specializing in water and wastewater treatment, biotechnology and bioengineering, fundamental chemical sciences, and river rehabilitation, particularly issues related to developing countries. Professor Din has received several awards and recognitions at both national and international levels.

List of Contributors

Benu Arora
Department of Applied Chemistry and Environmental Studies, Bhagwan Parshuram Institute of Technology, Guru Gobind Singh Indraprastha University, New Delhi, India.

Gagandeep Kaur
Chitkara School of Pharmacy, Chitkara University, Himachal Pradesh, India.

Mohd. Fadhil Md. Din
Center for Environmental Sustainability and Water Security (IPASA) & Department of Water and Environmental Engineering, School of Civil Engineering, Universiti Teknologi Malaysia, Bahru, Malaysia.

Neha Mudaliar
SVKM`s Mithibai College of Arts, Chauhan Institutes of Science & Amrutben Jivanlal College of Commerce and Economics, Mumbai, India.

Pratik Kale
K.J. Somaiya College, Mumbai, India.

Sandhya Mulchandani
Microbiology Department, Smt. Chandibai Himathmal Mansukhani College, Ulhasnagar, India.

Saraswathy Nagendran
SVKM`s Mithibai College of Arts, Chauhan Institutes of Science & Amrutben Jivanlal College of Commerce and Economics, Mumbai, India.

Sourabh Patil
School of Biotechnology, Vidya Pratishthan's Arts, Commerce and Science College, Baramati, Maharashtra, India.

Tushar Borse
School of Biotechnology, Vidya Pratishthan's Arts, Commerce and Science College, Baramati, Maharashtra, India.

Vikas Sharma
Guru Gobind Singh College of Pharmacy, Yamuna Nagar, Haryana, India.

Vasudeo Zambare
R&D Department, Balaji Enzyme and Chemical Pvt Ltd, Andheri (E), Mumbai, India &
Center for Environmental Sustainability and Water Security (IPASA) & Department of Water and Environmental Engineering, School of Civil Engineering, Universiti Teknologi Malaysia, Bahru, Malaysia.

Index

A

absorption, ix, xiii, xiv, 1, 2, 4, 5, 7, 8, 9, 10, 11, 12, 15, 16, 17, 19, 23, 25, 27, 28, 29, 31, 33, 34, 42, 43, 45, 46, 47, 49, 50, 51, 52, 54, 55, 56, 57, 58, 60, 105, 121, 123, 124, 129, 131, 132, 135, 153, 154, 155, 160
adipose tissue, 1, 2, 4, 5, 6, 7, 9, 18, 20, 21, 30, 31, 32, 34, 35, 47, 49, 123, 124, 125, 127, 132, 133, 143, 145, 147, 156, 160
advancements, 17, 35, 95
animals, x, 46, 65, 70, 72, 75, 76, 81, 92, 122, 125, 127, 129, 141, 143, 152, 153, 154, 156
anti-biofilm, 104
antimicrobial, x, 30, 68, 88, 89, 100, 102, 103, 104, 107, 108, 109, 110, 111, 113, 117, 119
anti-microbial activities, 88

B

biocatalyst, xiv, 65, 72, 78, 79, 88, 89, 97, 100, 101, 106, 111, 118, 134
biofuel, 164
bio-medical, 102
bio-surfactants, x, 87, 88, 89, 91, 92, 98, 102, 104, 106
biotechnology, 164, 165
birds, x, 141, 143, 154, 155, 156

C

cancer, x, xiii, 23, 29, 32, 40, 44, 49, 53, 56, 60, 61, 64, 68, 69, 80, 81, 83, 84, 88, 100, 105, 106, 112, 113, 118, 127, 133, 136, 144, 160
cardiovascular diseases, 13, 32, 33, 35, 80, 141, 142
cardiovascular health, 23, 24, 37, 39
catalysts, ix, 24, 65, 72, 77, 84, 89, 91, 108, 110, 122, 128, 138
cellular, ix, x, 1, 2, 3, 4, 8, 9, 10, 11, 17, 19, 21, 32, 35, 40, 61, 85, 121, 126, 128, 135, 139, 141, 145, 150, 153, 159
cholesterol, xiv, xv, 11, 14, 25, 26, 27, 28, 29, 30, 34, 35, 36, 37, 38, 42, 45, 46, 48, 55, 56, 57, 75, 80, 122, 123, 130, 132, 148, 149, 154, 159, 160, 161
classification, 1, 2, 3, 4, 5, 28, 60, 110, 117, 137, 161
conditions, x, 2, 3, 10, 12, 13, 14, 15, 17, 29, 32, 35, 36, 42, 44, 55, 57, 58, 59, 60, 68, 72, 74, 77, 80, 82, 83, 89, 92, 93, 103, 106, 111, 114, 123, 124, 129, 130, 132, 141, 142, 143, 152, 154, 156
cystic fibrosis, 15, 33, 49, 52, 55, 56, 60, 130, 142, 143, 149, 153, 156, 161

D

deep eutectic solvents (DESs), xiii, 93, 97, 99, 108, 110, 117
deficiency, 1, 3, 12, 13, 15, 17, 18, 19, 20, 30, 34, 123, 138, 145, 147, 154, 156, 158, 161, 162
developing countries, 165
diagnostic, 1, 2, 3, 13, 14, 15, 17, 59, 82, 121, 128, 131, 134, 152, 156, 158

Index

digestion, ix, x, 1, 2, 4, 5, 6, 7, 8, 12, 15, 17, 18, 21, 23, 25, 26, 27, 28, 31, 33, 49, 51, 52, 53, 54, 55, 56, 58, 59, 121, 124, 128, 129, 131, 132, 135, 137, 141, 143, 149, 151, 152, 153, 154, 155

digestive, ix, 2, 7, 8, 15, 20, 23, 25, 26, 43, 47, 49, 54, 55, 56, 57, 72, 83, 85, 122, 123, 128, 129, 130, 131, 134, 137, 138, 139, 144, 149, 151, 154, 159, 161

disease(s), ix, x, xi, xii, xiii, xiv, xv, 1, 2, 3, 5, 12, 13, 14, 15, 17, 18, 19, 21, 23, 28, 29, 30, 31, 32, 33, 34, 35, 36, 37, 38, 39, 40, 42, 43, 44, 49, 51, 53, 57, 60, 61, 62, 63, 64, 66, 67, 68, 70, 83, 115, 125, 128, 132, 134, 135, 137, 138, 139, 141, 143, 144, 145, 146, 147, 148, 149, 152, 153, 154, 155, 156, 157, 158, 159, 160, 161, 162

drug(s), x, xiv, xv, 17, 31, 37, 51, 52, 54, 56, 57, 60, 61, 63, 69, 72, 79, 80, 88, 104, 105, 114, 118, 119, 128, 131, 137, 141, 156

dysregulation, 10, 12, 13, 15, 16, 20, 28, 29, 31, 32

E

energy, x, xiii, 1, 2, 4, 6, 7, 8, 9, 10, 11, 17, 24, 29, 30, 31, 32, 34, 35, 37, 43, 44, 45, 48, 50, 67, 72, 76, 80, 82, 98, 121, 122, 125, 126, 128, 130, 132, 137, 143, 144, 145, 156

engineering, 165

environmental impact, 165

environmental management, 165

enzymatic, x, xi, 1, 2, 3, 4, 5, 8, 24, 25, 26, 35, 87, 89, 99, 106, 107, 108, 109, 110, 111, 112, 113, 114, 115, 117, 118, 119, 128, 131, 134, 150

enzyme(s), 163, 164

exploring, ix, x, 3, 5, 23, 31, 59, 124

F

fatty, x, xiv, xv, 1, 2, 4, 5, 6, 7, 8, 9, 10, 11, 12, 13, 18, 19, 23, 25, 26, 27, 30, 31, 32, 34, 35, 37, 43, 47, 48, 54, 56, 72, 74, 78, 80, 81, 87, 88, 89, 91, 93, 94, 95, 96, 97, 99, 104, 107, 108, 109, 110, 112, 114, 115, 117, 118, 119, 121, 122, 123, 124,125, 126, 127, 128, 129, 130, 141, 142, 143, 144, 150, 155, 156, 158, 160, 161

fatty acid, x, 1, 2, 4, 5, 6, 7, 8, 9, 10, 11, 12, 13, 23, 25, 26, 27, 30, 31, 32, 34, 35, 37, 43, 47, 48, 54, 56, 72, 74, 78, 80, 81, 87, 89, 91, 93, 94, 95, 96, 97, 99, 104, 119, 121, 122, 123, 124, 125, 126, 127, 128, 129, 130, 141, 142, 144, 150, 156, 161

fatty acid vinyl esters (FAVEs), 93, 97, 99, 100, 107

fermentation, 163, 164

formation, 1, 7, 8, 10, 19, 21, 28, 34, 36, 37, 38, 48, 52, 54, 74, 81, 96, 97, 101, 104, 123, 146, 148, 149, 154

G

gastric, 1, 2, 4, 5, 6, 7, 10, 21, 25, 30, 46, 48, 51, 52, 54, 63, 127, 128, 130, 132, 143, 150

gene therapy, 16, 17, 53

glycolipids, x, 87, 88, 89, 90, 91, 92, 99, 102, 103, 104, 105, 106, 112, 117, 129

guardians, x, 121

H

health, ix, x, xi, xii, xv, 1, 2, 3, 5, 10, 11, 12, 13, 14, 15, 17, 18, 23, 24, 28, 30, 31, 33, 35, 36, 38, 43, 44, 45, 47, 55, 59, 62, 63, 66, 67, 68, 69, 70, 79, 80, 81, 83, 84, 87, 88, 90, 102, 107, 108, 129, 141, 146, 154, 161

health benefits, 23, 24, 31, 66, 67, 68, 70, 80, 81

hepatic, xiv, 20, 21, 25, 36, 37, 47, 60, 64, 75, 85, 123, 127, 135, 138, 141, 142, 144, 148, 154, 155, 156, 157, 161
high fat diet, 122
homeostasis, 1, 2, 4, 6, 7, 9, 10, 13, 17, 35, 39, 40, 67, 125, 126, 139
humans, x, 60, 67, 68, 70, 123, 125, 128, 141, 143, 144, 145, 147, 150, 152, 153, 154, 156
hydrolysis, 1, 2, 4, 5, 6, 7, 8, 10, 11, 12, 19, 25, 29, 30, 32, 34, 42, 47, 49, 62, 66, 74, 80, 81, 83, 92, 122, 123, 126, 127, 129, 130, 131, 133, 134, 142, 143, 144, 161

I

imaging, xv, 1, 3, 13, 14, 15, 17, 53, 152
immune system, ix, 1, 3, 10, 11, 30, 39, 68
industries, ix, x, 65, 70, 73, 75, 79, 80, 82, 84, 87, 106, 108, 114, 121, 122, 133, 134
inflammation, 12, 14, 15, 19, 24, 29, 30, 32, 33, 36, 37, 39, 40, 41, 42, 43, 47, 50, 52, 61, 63, 64, 68, 69, 125, 126, 135, 136, 138, 141, 143, 149, 150, 151, 157, 159, 160
inhibition, ix, 17, 19, 40, 41, 42, 43, 49, 62, 63, 65, 68, 72, 73, 103, 105, 132, 133, 137, 141
innovations, ix, 17, 65
insufficiency, xiv, xv, 15, 17, 23, 33, 49, 50, 51, 53, 55, 58, 59, 60, 61, 63, 130, 136, 139, 144, 148, 150, 153, 154, 157, 158, 161
insulin, 1, 2, 6, 7, 9, 10, 13, 19, 20, 21, 29, 32, 33, 35, 36, 37, 45, 46, 47, 80, 125, 136, 143, 145, 160

L

lipid, ix, x, xiv, xv, 1, 2, 3, 4, 5, 6, 7, 8, 9, 10, 11, 13, 14, 15, 17, 18, 19, 20, 21, 23, 26, 27, 28, 29, 31, 32, 33, 34, 35, 36, 38, 39, 42, 43, 47, 48, 59, 60, 61, 62, 63, 64, 72, 78, 103, 110, 113, 121, 125, 126, 127, 130, 131, 132, 133, 135, 136, 137, 138, 139, 141, 142, 143, 144, 145, 146, 147, 148, 151, 152, 154, 156, 159, 162
lipoprotein(s), xiv, xv, 1, 2, 4, 5, 6, 7, 8, 18, 19, 20, 21, 25, 28, 32, 36, 37, 38, 39, 41, 48, 60, 61, 62, 63, 65, 75, 80, 85, 121, 123, 124, 133, 135, 137, 138, 139, 141, 142, 144, 145, 146, 147, 148, 154, 156, 157, 158, 159, 160, 161
liver, 8, 19, 25, 29, 31, 36, 38, 40, 42, 47, 57, 62, 63, 74, 105, 123, 125, 127, 128, 133, 138, 141, 142, 147, 148, 149, 151, 153, 154, 155, 156, 158, 159, 160, 161
low-density lipoprotein, 28, 38, 46, 75, 122, 123, 142, 146, 147

M

malabsorption syndrome, 153, 155, 160
mechanism, 2, 4, 30, 34, 39, 42, 43, 47, 68, 69, 111, 132, 134, 138, 145, 150
metabolic disorder, 1, 3, 10, 12, 13, 14, 16, 28, 29, 31, 32, 33, 60, 124, 145
metabolic influence, ix, 23
metabolism, ix, x, 1, 2, 3, 5, 6, 7, 9, 10, 11, 13, 17, 18, 19, 20, 21, 23, 24, 28, 29, 31, 32, 33, 34, 35, 36, 37, 38, 39, 43, 45, 47, 60, 61, 62, 63, 64, 75, 81, 84, 85, 105, 121, 123, 126, 127, 128, 133, 134, 135, 136, 137, 138, 139, 142, 144, 145, 147, 148, 152, 154, 156, 157, 159, 161
micelle, xiii, 8, 19, 52, 54, 91
microbial, x, 18, 33, 60, 62, 66, 71, 75, 76, 77, 79, 82, 85, 87, 89, 92, 100, 104, 107, 108, 109, 114, 117, 118, 132, 134, 136, 137, 161
microbial lipases, 66, 75, 76, 79
modulation, ix, 1, 9, 10, 11, 17, 19, 31, 42, 43, 68, 116, 133, 134
molecular, ix, x, xv, 1, 2, 3, 5, 8, 17, 19, 20, 46, 60, 62, 63, 73, 76, 78, 82, 92, 101, 103, 109, 110, 111, 115, 116, 118, 127, 137, 139, 145, 148, 158, 159, 161

Index

N

nucleotide sequence, 164
nutraceuticals, 66, 72, 80
nutrient, ix, 1, 2, 4, 7, 9, 10, 11, 15, 17, 58, 72, 83, 85, 144, 145, 160

O

obesity, ix, 10, 13, 16, 17, 18, 20, 21, 28, 29, 31, 32, 33, 35, 36, 40, 44, 45, 47, 60, 62, 63, 65, 68, 80, 84, 124, 132, 134, 138, 141, 142, 143, 145, 155, 158, 160, 161
organic solvents, 78, 88, 95, 97, 98, 100, 101

P

pancreas, 4, 5, 7, 9, 12, 14, 15, 18, 23, 24, 26, 31, 47, 49, 54, 56, 60, 62, 105, 122, 123, 124, 130, 133, 137, 143, 149, 151, 153, 154, 160
pancreatic, ix, xiv, xv, 1, 2, 4, 5, 6, 7, 10, 12, 13, 14, 15, 16, 17, 18, 19, 21, 23, 25, 26, 29, 30, 33, 42, 46, 47, 48, 49, 50, 51, 52, 53, 55, 56, 58, 59, 60, 61, 62, 63, 65, 74, 81, 84, 122, 124, 127, 128, 129, 130, 132, 133, 134, 136, 137, 138, 139, 141, 142, 144, 149, 150, 151, 152, 153, 154, 156, 157, 158, 161
pancreatitis, xiii, 1, 3, 12, 13, 14, 15, 18, 20, 25, 49, 55, 56, 60, 128, 130, 132, 137, 139, 143, 144, 150, 151, 152, 153, 156, 157, 158, 159, 161, 162
pharmaceutical, 45, 59, 80, 87, 110, 115, 128, 130, 131, 132, 134, 137, 141
physiological, ix, x, 2, 5, 7, 10, 12, 17, 18, 19, 28, 29, 35, 53, 121, 122, 128, 143, 149, 154, 158
plant-based, 132
probiotic isolates, 65, 66
probiotics, 66, 67, 68, 69, 70, 71, 72, 81, 83, 84, 85, 124, 129, 164

R

rehabilitation, 165
replacement, xiv, xv, 2, 3, 15, 17, 18, 34, 51, 52, 60, 61, 136, 144, 157

S

structure(s), ix, 1, 2, 3, 5, 8, 14, 20, 21, 27, 37, 42, 46, 65, 73, 79, 90, 91, 92, 99, 107, 109, 110, 114, 115, 116, 118, 122, 134, 139, 149, 159
sugar, xv, 45, 87, 88, 89, 91, 93, 94, 97, 98, 100, 101, 107, 109, 110, 112, 114, 115, 117, 118
surfactants, x, 78, 82, 87, 88, 90, 91, 103, 104, 106, 108, 109, 110, 111, 114, 115, 117, 118
sustainability, 165
sustainable, ix, 65, 78, 81, 99, 107, 109, 110, 113, 115, 117
synthesis, x, 11, 31, 35, 38, 49, 72, 74, 78, 80, 83, 84, 85, 87, 89, 90, 91, 92, 93, 95, 96, 97, 99, 101, 106, 107, 108, 109, 110, 111, 112, 113, 114, 115, 116, 117, 118, 119, 125, 127, 131, 134, 147, 148

T

therapeutic, ix, x, xv, 2, 3, 10, 13, 15, 16, 17, 18, 19, 23, 29, 31, 33, 35, 39, 53, 58, 62, 63, 65, 82, 88, 102, 105, 108, 121, 124, 128, 131, 134, 136, 137, 141, 147, 157, 158
therapeutic agents, 29, 88, 124
transportation, 165
treatment, 16, 29, 30, 31, 35, 40, 42, 45, 46, 51, 53, 59, 60, 61, 62, 63, 69, 81, 83, 87, 105, 128, 130, 131, 132, 133, 136, 144, 157, 158, 161, 165
triacylglyceride, 125
triglycerides, x, xv, 2, 4, 5, 6, 7, 8, 9, 10, 14, 23, 25, 26, 27, 30, 31, 32, 35, 36, 38, 42, 47, 48, 54, 55, 81, 121, 122, 123,

Index

126, 127, 128, 131, 133, 142, 143, 144, 147, 148, 154

W

waste management, 164

wastewater, 165
weight management, 16, 23, 24, 43, 44, 45, 47